BRILL Multisensory Research 26 Supplement (2013) 1–2

Abstracts from the 14th International Multisensory Research Forum
The Hebrew University in Jerusalem, June 3rd–6th 2013

It gives us great pleasure to welcome you to the 14th Annual IMRF Meeting in Jerusalem, Israel, which promises to be an exciting and dynamic conference. We gathered here from June 3–6, 2013 at the Dan Hotel overlooking the old city of Jerusalem, and embracing the cutting edge of multisensory research. This conference has travelled the world — last year we returned to Oxford England where the first conference was first held in 1999; this year we again start moving around the globe visiting as many new places as we have new research results.

The abstracts included in this supplement to the journal *Multisensory Research* constitute the symposia, talks and posters presented at IMRF 2013. The broad interest in the field of multisensory perception, as well as its applicability in many aspects of our lives, are demonstrated by the wide variety of subjects covered in these abstracts. The wide international appeal is evident in the diversity of countries represented in the forum. The research presented uses a wide range of methods including: psychophysics, neuropsychology, neuroimaging, neural stimulation, and computational modeling. This year's conference maintains the high standards typical of the forum.

In terms of student awards, we received close to thirty applications, and had a very difficult time deciding on the top candidates — all of the submissions were of very high quality. The top 5 ranked abstracts are noted in the schedule; their talks are interlaced with the regular-session talks. The high quality of the remaining student applications warranted their inclusion in a special student-only session. While this is different from previous IMRF conferences, we believe it highlights the exceptional trainees and mentors in our community.

In this book, you will find the abstracts in alphabetical order by the first author, within each of three sections: symposia, talks and posters. We would

DOI:10.1163/22134808-000S0001

like to thank Brill Publishing for producing this volume, and for doing it on such a short timeline. It has been a pleasure working with their team.

As in previous years, there will be a call for papers to be published in a Special Issue. This year, as for the 2012 conference, we have selected the journal *Multisensory Research* for this publication (see Call for Papers on the last page).

Finally, we would like to thank the Advisory Board members who contributed their time to help select the symposia and student award winners for this year: Yoram Gutfreund, Vanessa Harrar, Micah M. Murray, Fiona N. Newell, Daniel-Robert Chebat, Miriam Reiner, Krish Sathian, Charles Spence, Jean Vroomen, and Mark T. Wallace.

We trust you will enjoy this year's IMRF in Jerusalem, where 5000 years of history juxtaposes modern life, and where variety is the name of the game.

Prof. Amir Amedi
Department of Medical Neurobiology
Institute for Medical Research
Israel–Canada (IMRIC) &
The Edmond and Lily Safra Center
for Brain Sciences (ELSC)
& The Cognitive Science Program
The Hebrew University of Jerusalem
Israel

Prof. David I. Shore
Department of Psychology,
Neuroscience & Behaviour
McMaster University, Canada

Dr. Shelly Levy-Tzedek
Department of Medical Neurobiology
Institute for Medical Research
Israel–Canada (IMRIC) &
The Edmond and Lily Safra Center for
Brain Sciences (ELSC)
The Hebrew University of Jerusalem
Israel

BRILL Multisensory Research 26 Supplement (2013) 3 brill.com/msr

Keynote Presentation

Processing streams in the early blind

Josef P. Rauschecker *

Georgetown University Medical Center, Washington, DC, USA

Abstract

It has long been established that the occipital cortex of early blind animals and humans is activated by nonvisual stimuli (Rauschecker, 1995). This has led to the concept of cross-modal compensatory plasticity. Using functional magnetic resonance imaging we have demonstrated that the middle occipital gyrus (MOG) of early blind humans retains its function in spatial localization and is activated in auditory and tactile spatial tasks (Renier *et al.*, 2010). At the same time, visual-to-auditory substitution devices have been used to demonstrate that cortical areas in the visual ventral stream retain their function in object recognition, when they are stimulated by auditory and tactile 'shapes'. The results confirm that cross-modal plasticity can restitute perceptual abilities in the blind that resemble vision. These abilities are supported by activating specialized processing streams in normally 'visual' cortex *via* preserved nonvisual inputs.

References

Anurova, I., Renier, L. A., De Volder, A. G., Carlson, S. and Rauschecker, J. P. (2013). Cortical structure and function in the early blind (in prep.).

Rauschecker, J. P. (1995). Compensatory plasticity and sensory substitution in the cerebral cortex, *Trends in Neurosci.* **18**, 36–43.

Renier, L. A., Anurova, I., De Volder, A. G., Carlson, S., VanMeter, J. and Rauschecker, J. P. (2010). Preserved functional specialization for spatial processing in the occipital cortex of the early blind, *Neuron* **68**, 138–148.

Abstract from the 14th International Multisensory Research Forum, The Hebrew University of Jerusalem, Israel, 2013.

* Contact author: rauschej@georgetown.edu

 DOI:10.1163/22134808-000S0002

Section I: SYMPOSIA

Symposium #1:
Spatial transformations in multisensory integration

Stephanie Badde (Chair), Elena Azanon, Shigeru Kitazawa, Jared Medina and David I. Shore

BRILL Multisensory Research 26 Supplement (2013) 8 brill.com/msr

Talk, "Spatial transformations in multisensory integration" Symposium

Modeling body posture effects on reference frame integration

Stephanie Badde *, **Tobias Heed and Brigitte Roeder**

Biological Psychology and Neuropsychology, University of Hamburg, Germany

Abstract

To optimally estimate the location of a stimulus, the brain integrates spatial information from different senses. For such integration, spatial information available in modality-specific reference frames must be transformed into a common reference frame. For example, touch is remapped from skin-based, anatomical coordinates into an external, visual reference frame.

Touch remapping is usually investigated by setting anatomical and external coordinates into conflict. For example, many studies have used temporal order judgments (TOJ) of tactile stimuli, for which performance is markedly impaired when the hands are crossed.

Here, we show that crossing effects are not a peculiarity of the TOJ task, and demonstrate crossing effects in several tactile localization tasks. The size of the crossing effect was strongly related across tasks within participants, suggesting that crossing effects in different tasks originated from a common process. Furthermore, crossing effects were systematically modulated by both bottom-up and top-down features. Furthermore, we demonstrate that the data from all investigated tasks can be accounted for by a generative, probabilistic model of touch localization. The core feature of this model is the estimation of stimulus location by a weighted combination of the different reference frames. This result argues against the idea that crossing effects may be due to failures of the remapping calculations from skin to external space. Rather, the model suggests that crossing effects are a result of the brain's attempt to integrate every available piece of information, independent of current posture.

Keywords

Remapping, tactile, probabilistic model

Abstract from the 14th International Multisensory Research Forum, The Hebrew University of Jerusalem, Israel, 2013.

* Contact author: stephanie.badde@uni-hamburg.de

DOI:10.1163/22134808-000S0003

BRILL Multisensory Research 26 Supplement (2013) 9–10 brill.com/msr

Talk, "Spatial transformations in multisensory integration" Symposium

Assessing the effects of posture changes in tactile remapping

Elena Azanon [1,*], **Flavia Cardini** [2] **and Patrick Haggard** [1]

[1] Institute Cognitive Neuroscience, University College London, UK
[2] Royal Holloway, University of London, London, UK

Abstract

During the apparently simple act of localizing a tactile event on the skin, the brain must realign spatial representations according to current body posture, from somatotopic (skin based) to externally based coordinates. Previous studies have highlighted the obligatory nature of this process which seems to be carried out every time a tactile event is felt, even when posture remains constant (Yamamoto and Kitazawa, 2001). Nonetheless, the way this system is modulated by continued adaptation to a static posture is still largely unknown. Here, we address the way posture changes affect the accuracy in which touch is remapped. Participants in our experiment were asked to compare the location of two tactile stimuli presented to either hand when posture of the limbs was frequently changed (crossed *vs.* uncrossed) or else, remained constant throughout an entire block. We found that frequent changes induces more errors when localizing touch using external space, even though proprioceptive signal decreases with time. Furthermore, the advantage of keeping the arms in a constant posture was observed in a trial by trial basis: in the interleaved condition, participants' responses were more precise when the previous trial had been performed in the same position. Previous results suggest that remapping might take place using a canonical configuration of the hands (Azanon and Soto-Faraco, 2008; Yamamoto and Kitazawa, 2001). If this were the case, then a change of posture might require a new 'remapping model' to be activated, which might require some time. These results suggest that the remapping system can be quickly adapted, suggesting a certain amount of short term plasticity that allows for an adaptive use of postural information.

Keywords

Tactile remapping, proprioception, adaptation

Abstract from the 14th International Multisensory Research Forum, The Hebrew University of Jerusalem, Israel, 2013.

* Contact author: eazanyon@gmail.com

 DOI:10.1163/22134808-000S0004

References

Azanon, E. and Soto-Faraco, S. (2008). Changing reference frames during the encoding of tactile events, *Curr. Biol.* **18**, 1044–1049.

Yamamoto, S. and Kitazawa, S. (2001). Reversal of subjective temporal order due to arm crossing, *Nat. Neurosci.* **4**, 759–765.

BRILL Multisensory Research 26 Supplement (2013) 11 brill.com/msr

Talk, "Spatial transformations in multisensory integration" Symposium

Where tactile signals are ordered in time

Shigeru Kitazawa *

Graduate School of Frontier Biosciences, Osaka University, Japan

Abstract

We previously found that crossing the arms caused misreporting of the temporal order of successive tactile stimuli, delivered one to each hand (Yamamoto and Kitazawa, 2001). From the results, we suggested that it is not until the tactile signals are localized in space that the signals are ordered in time. It is now generally accepted that a tactile stimulus delivered to a hand is initially mapped to the wrong hand, and then remapped to the correct hand, when the arms are crossed. This idea derived from a phenomenon termed 'curved somatosensory saccade' (Groh and Sparks, 1996; Overvliet *et al.*, 2011). We hypothesized that (1) an 'inverted' motion vector is generated from the wrong mapping of the first stimulus to the wrong mapping of the second stimulus, and (2) reversal occurs because the order of tactile events are reconstructed by combining information regarding 'what happened where' and information regarding 'when', captured in an inverted motion vector (Takahashi *et al.*, 2012). We provide evidence that supports the 'motion projection' hypothesis, and show that the process including remapping takes place in areas that receive converging inputs from tactile, proprioceptive, and visual modalities, as well as corollary discharges of saccade control signals.

Keywords

Spatial remapping, temporal order judgment, fMRI

References

Groh, J. M. and Sparks, D. L. (1996). Saccades to somatosensory targets. I. Behavioral characteristics, *J. Neurophysiol.* **75**, 412–427.

Overvliet, K. E., Azanon, E. and Soto-Faraco, S. (2011). Somatosensory saccades reveal the timing of tactile spatial remapping, *Neuropsychologia* **49**, 3046–3052.

Takahashi, T., Kansaku, K., Wada, M., Shibuya, S. and Kitazawa, S. (2012). Neural correlates of tactile temporal-order judgment in humans: an fMRI study, *Cereb. Cortex*.

Yamamoto, S. and Kitazawa, S. (2001). Reversal of subjective temporal order due to arm crossing, *Nat. Neurosci.* **4**, 759–765.

Abstract from the 14th International Multisensory Research Forum, The Hebrew University of Jerusalem, Israel, 2013.

* Contact author: kitazawa@fbs.osaka-u.ac.jp

 DOI:10.1163/22134808-000S0005

BRILL *Multisensory Research 26 Supplement (2013) 12* MULTISENSORY RESEARCH brill.com/msr

Talk, "Spatial transformations in multisensory integration" Symposium

Dynamic shifts in tactile localization after stroke

Jared Medina *

University of Delaware, USA

Abstract

Plasticity in somatosensory maps has been observed after brain damage and differential stimulation. These studies demonstrate that the relationship between tactile stimulation and activation in primary somatosensory maps is labile. However, few studies have examined the perceptual consequences of somatosensory reorganization in humans. Repeated tactile stimulation can result in rapid changes in both somatosensory map topography and tactile perception in neurologically-intact individuals. Given evidence for rapid activity-dependent plasticity of somatosensory maps, along with the need to incorporate higher-order representations of the body to localize touch, we hypothesized that tactile localization could be shifted based on the location of previous stimulation in brain-damaged individuals. Therefore, we instructed two individuals with left primary somatosensory damage to localize suprathreshold tactile stimuli presented to the hand only, or to the hand preceded by forearm stimulation. On hand only trials, we found systematic mislocalization of tactile sensation, likely due to a mismapping from primary somatosensory maps to a higher-order representation of body size and shape. However, when preceded by forearm stimulation, localization judgments of hand stimuli shifted proximally on the contralesional, but not ipsilesional, limb. In an additional experiment, we presented stimuli to the medial segment of the contralesional fingers, and found that localization judgments were shifted in the direction of prior stimulation on the distal or proximal segment of the same finger. Our results, consistent with decreased inhibition in sensory cortex after stroke, may reflect rapid changes in primary somatosensory cortex combined with a continued mismapping from sensory maps to higher-order body representations.

Keywords

Body representations, plasticity, somatosensory cortex

Abstract from the 14th International Multisensory Research Forum, The Hebrew University of Jerusalem, Israel, 2013.

* Contact author: jmedina@psych.udel.edu

 DOI:10.1163/22134808-000S0006

BRILL Multisensory Research 26 Supplement (2013) 13 brill.com/msr

Talk, "Spatial transformations in multisensory integration" Symposium

Reference frame conflict: Blindfolding and the crossed-hands deficit

David I. Shore * and Michelle L. Cadieux

Multisensory Perception Laboratory, McMaster University, Hamilton, Ontario, Canada

Abstract

The crossed-hands temporal order judgment deficit highlights the automatic remapping of tactile stimuli from a body-surface representation into a body-independent representation. In considering the source of the deficit two potentially complementary ideas have emerged — the transformation could be slowed by the awkward posture (Yamamoto and Kitazawa, 2001) or it may take time to resolve the conflict between the internal and external frames of reference (Shore *et al.*, 2002). We manipulated the availability of visual information across four experiments varying in their response demands. When observers responded with the hand that was stimulated (as opposed to with the foot pedals) there was a significant reduction in the size of the crossed-hands deficit when vision was prevented by blindfolding. These data support a conflict resolution account of the deficit over a slower remapping account.

Keywords

Visuotactile, blindfolding, crossed-hands, temporal order judgment

References

Shore, D. I., Spry, E. and Spence, C. (2002). Confusing the mind by crossing the hands, *Cogn. Brain Res.* **14**, 153–163.

Yamamoto, S. and Kitazawa, S. (2001). Reversal of subjective temporal order due to arm crossing, *Nat. Neurosci.* **4**, 759–765.

Abstract from the 14th International Multisensory Research Forum, The Hebrew University of Jerusalem, Israel, 2013.

* Contact author: dshore@mcmaster.ca

DOI:10.1163/22134808-000S0007

Symposium #2:
Contemporary multisensory research in animal models

Yoram Gutfreund (Chair), Adi Mizrahi, Rony Paz and Nachum Ulanovsky

BRILL Multisensory Research 26 Supplement (2013) 16 brill.com/msr

Talk, "Contemporary multisensory research in animal models" Symposium

Visual and auditory processing for target selection

Yoram Gutfreund *

Department of Physiology and Biophysics, the Technion, Haifa, Israel

Abstract

The barn owl (*Tyto alba*) evolved precise visual and auditory systems to detect small prey in acoustically noisy and dimly lit conditions. Consequently, this species provides an excellent model system for studying visual–auditory integration for target selection. We study neural responses to visual and auditory stimuli in the optic tectum (OT), a mid-brain area where sensory information is integrated to guide the body responses to salient stimuli. I will focus my talk on neural adaptation, the decline of the neural response to repetitive stimulation. Adaptation in the OT is ubiquitous and robust. A single short stimulus can significantly reduce the neural responses to subsequent stimuli appearing up to 60 second later. The adaptation is specific to the stimulus, changing the subsequent stimulus (frequency, amplitude, location, etc.) regains the neural response. Thus, tectal neurons are sensitive to changes in the input stream, with a memory trace of up to a minute and possibly more. This sensitivity to novel events is enhanced when visual and auditory stimuli are presented simultaneously from the same location. We propose, and provide evidence, that the multimodal adaptation in the OT reflects the neural computation that is required for behavioral habituation of the orientation reflex.

Keywords

Attention, superior colliculus, barn owl, auditory localization

Abstract from the 14th International Multisensory Research Forum, The Hebrew University of Jerusalem, Israel, 2013.

* Contact author: yoramgut@gmail.com

 DOI:10.1163/22134808-000S0008

BRILL Multisensory Research 26 Supplement (2013) 17 brill.com/msr

Talk, "Contemporary multisensory research in animal models" Symposium

Natural odors modulate processing of sounds in motherhood

Adi Mizrahi *

Department of Neurobiology, Inst. for Life Sciences and the Edmond and Lily Safra Center for Brain Sciences, the Hebrew University of Jerusalem, Israel

Abstract

Motherhood is associated with different forms of physiological alterations including transient hormonal changes and brain plasticity. The underlying impact of these changes on the emergence of maternal behaviors and sensory processing within the mother's brain are largely unknown. I will discuss our recent work using electrophysiology in lactating mice shortly after giving birth to their first litter. Using *in vivo* cell-attached recordings in the primary auditory cortex (A1) we found that lactating mothers showed a novel form of multisensory interaction. Specifically, exposure to pups body odors reshaped neuronal responses to auditory stimuli in A1. This novel olfactory–auditory interaction appeared naturally in lactating mothers shortly after parturition and was long lasting. Naive virgins that had experience with the pups also showed an appearance of olfactory–auditory integration in A1, suggesting that this phenomenon may be experience dependent. I will discuss our recent data using targeted two photon patch clamp to study how multisensory plasticity is represented by specific populations of interneurons. Our data reveals that plasticity is tunneled *via* decreases in feed-forward inhibitory circuits.

Abstract from the 14th International Multisensory Research Forum, The Hebrew University of Jerusalem, Israel, 2013.

* Contact author: mizrahi.adi@mail.huji.ac.il

 DOI:10.1163/22134808-000S0009

BRILL Multisensory Research 26 Supplement (2013) 18 brill.com/msr

Talk, "Contemporary multisensory research in animal models" Symposium

Modality-invariant mechanisms of aversive learning: Behavior and brain mechanisms

Rony Paz *

Department of Neurobiology, Weizmann Institute, Israel

Abstract

Risks to survival require similar responses, independent of the sense they were perceived with. Generalization of learning is one such example, where a 'better safe than sorry' rule should govern. I will discuss how such response is achieved *via* perception and/or choice-behavior, and what brain mechanisms can underlie the sensory-invariant effect. Results from psychophysics, human imaging, and electrophysiology in animal models will be presented to shed light on the network architecture that supports this.

Abstract from the 14th International Multisensory Research Forum, The Hebrew University of Jerusalem, Israel, 2013.

* Contact author: rony.paz@weizmann.ac.il

 DOI:10.1163/22134808-000S0010

BRILL Multisensory Research 26 Supplement (2013) 19 brill.com/msr

Talk, "Contemporary multisensory research in animal models" Symposium

Multi-sensory determinants of 3D spatial codes in the hippocampal formation of bats

Nachum Ulanovsky *

Department of Neurobiology, Weizmann Institute, Israel

Abstract

The hippocampus, although not typically considered a multi-sensory hub, is in fact receiving rich converging inputs from all sensory modalities. We study the neural basis of spatial behaviors and spatial cognition in freely-moving, freely behaving mammals — employing the echolocating bat as a novel animal model. I will describe our recent studies of 3D place cells and 3D head-direction cells in crawling and flying bats, and will describe the effects of trading-off sensory inputs (echolocation *versus* vision).

Abstract from the 14th International Multisensory Research Forum, The Hebrew University of Jerusalem, Israel, 2013.

* Contact author: nachum.ulanovsky@weizmann.ac.il

 DOI:10.1163/22134808-000S0011

Symposium #3: Atypical multisensory integration

Vanessa Harrar (Chair), Andrew J. Bremner, Carissa Cascio and Ryan Stevenson

BRILL Multisensory Research 26 Supplement (2013) 22 brill.com/msr

Talk, "Atypical multisensory integration" Symposium

Multisensory integration deficits in developmental dyslexia

Vanessa Harrar [1,*], Jonathan Tammam [2], Alexis Pérez-Bellido [3], Rachel Hulatt [4], Anna Pitt [2], John Stein [2] and Charles Spence [1]

[1] Department of Experimental Psychology, University of Oxford, GB
[2] Department of Physiology, Anatomy, and Genetics, University of Oxford, GB
[3] Dept. de Tecnologies de la Informació i les Comunicacions, Universitat Pompeu Fabra, ES
[4] Oriel College, University of Oxford, GB

Abstract

There is growing experimental support for the presence of specific deficits in the magnocellular visual pathway in dyslexia. The magnocellular system is thought to be involved in multisensory integration. As a result of impaired magnocellular function, specific deficits in multisensory integration may be observed in dyslexia. Multisensory integration differences were compared between dyslexics and matched controls, using simple reaction times. Four visual stimuli, preferentially activating the magnocellular or parvocellular visual systems, and white noise bursts were presented either alone or together and participants were instructed to respond as quickly as possible. Reaction times (RTs) to multisensory stimuli were predicted from the unisensory stimuli using Miller's Race model (which assumes independence of the two channels). RTs to multisensory stimuli that exceeded the model (faster than predicted) provided evidence of multisensory integration. Dyslexics integrated less than matched controls. Differences between groups were more pronounced for magnocellular-preferred stimuli, and literacy scores were significantly correlated with the range of RTs demonstrating integration. These results provide further support for the magnocellular theory of dyslexia, and suggest a simple non-literacy based diagnostic or experimental tools.

Keywords

Development, atypical integration, dyslexia

Abstract from the 14th International Multisensory Research Forum, The Hebrew University of Jerusalem, Israel, 2013.

* Contact author: vanessa.harrar@psy.ox.ac.uk

 DOI:10.1163/22134808-000S0012

BRILL Multisensory Research 26 Supplement (2013) 23 brill.com/msr

Talk, "Atypical multisensory integration" Symposium

Considering the development of developmental disorders of multisensory processes

Andrew J. Bremner * **and Elisabeth L. Hill**

Goldsmiths, University of London, UK

Abstract

By definition, developmental disorders of the brain and nervous system have their onset during early development. This early onset has important connotations for how we can characterize a disorder and its causal origins; the phenotypical outcome will be the result of an interaction between the initial atypical state of the individual and the subsequent ontogenetic developmental process. Despite patchy consideration given to the role of aberrant multisensory processes in developmental disorders, we are witnessing an increasing number of observations of multisensory impairments across a range of disorders, as exemplified by the other papers in this symposium. This paper discusses the current state of knowledge about aberrant multisensory processes across three developmental disorders: developmental coordination disorder, autism spectrum disorder, and dyslexia. Across these three disorders, distinct similarities in sensory responses are noted (e.g., hypo- and hyper-sensitivity to sensory information, sensorimotor impairments). This leads to a discussion of whether multisensory processing abnormalities could represent a particular vulnerability or, perhaps more importantly, a particular risk factor in atypical development. Within and across disorders, possible developmental trajectories that may have led to such impairments are examined. Finally, the need for further multisensory research in developmental disorders is highlighted and avenues for future research are explored.

Keywords

Multisensory, development, developmental disorders, autism spectrum disorder, developmental coordination disorder, dyslexia

Abstract from the 14th International Multisensory Research Forum, The Hebrew University of Jerusalem, Israel, 2013.

* Contact author: a.bremner@gold.ac.uk

DOI:10.1163/22134808-000S0013

BRILL Multisensory Research 26 Supplement (2013) 24 brill.com/msr

Talk, "Atypical multisensory integration" Symposium

Delayed influence of tactile–visual input on proprioception in autism: Evidence from the rubber hand illusion

Carissa Cascio *

Vanderbilt University, USA

Abstract

In the rubber hand illusion, perceived hand ownership can be transferred to a rubber hand after synchronous visual and tactile stimulation. Perceived body ownership and self-other relation are foundational for development of self-awareness, imitation, and empathy, which are all affected in autism spectrum disorders (ASD). We examined the rubber hand illusion in children with and without ASD. Children with ASD were initially less susceptible to the illusion than the comparison group, yet exhibited the effects of the illusion as time went on. Delayed susceptibility to the illusion may result from atypical multisensory temporal integration and/or an unusually strong reliance on proprioception. Children with ASD who display less empathy were significantly less likely to experience the illusion than those with more intact ability to express empathy. A better understanding of body representation in ASD may elucidate neural underpinnings of social deficits, thus informing future intervention approaches.

Keywords

Tactile, proprioception, autism

Abstract from the 14th International Multisensory Research Forum, The Hebrew University of Jerusalem, Israel, 2013.

* Contact author: carissa.cascio@vanderbilt.edu

 DOI:10.1163/22134808-000S0014

BRILL Multisensory Research 26 Supplement (2013) 25 brill.com/msr

Talk, "Atypical multisensory integration" Symposium

Atypical multisensory integration in Autism Spectrum Disorders: Cascading impacts of altered temporal processing

Ryan A. Stevenson [1,*], **Justin K. Siemann** [2], **Stephanie T. Brown** [3], **Tiffany G. Woynaroski** [2], **Magali Segers** [3], **James Bebko** [3] **and Mark Wallace** [2]

[1] Vanderbilt University Medical Center, Tennessee, USA
[2] Vanderbilt Brain Institute, Tennessee, USA
[3] York University, Toronto, Canada

Abstract

A strong factor influencing multisensory integration is the temporal relationship between the sensory inputs that are combined. Individuals with Autism Spectrum Disorders (ASD) exhibit both atypical multisensory and temporal processing deficits relative to their typically developing (TD) peers. A series of behavioral and fMRI studies from our lab have focused on the link between these two processes. Using speech and non-speech stimuli with parametrically varied temporal relationships between the auditory and visual components, we showed that multisensory temporal processing is indeed altered in ASD, with the largest deficits observed with speech stimuli. The temporal changes seen with simple, non-speech stimuli are strongly correlated with behavioral measures of perceptual binding of audiovisual speech, which suggests that low-level multisensory temporal deficits have cascading effects on speech perception.

To explore the neural substrates of these behavioral effects, we implemented an fMRI paradigm in individuals with ASD and TD where we presented synchronous and asynchronous speech and non-speech stimuli. We functionally localized a region in the superior temporal sulcus (pSTS), based on its involvement in multisensory binding and temporal processing and known functional and anatomical differences in ASD. Responses to audiovisual stimuli were extracted and compared across stimulus types and groups. Both TD and ASD groups show reduced pSTS activation with synchronous relative to asynchronous non-speech presentations, reflecting increased processing efficiency. For speech stimuli, only the TD group showed this effect. These data suggest differences in neural processing in pSTS may be at the core of atypical speech perception observed in ASD.

Keywords

Autism, speech, audiovisual, temporal, timing, multisensory, superior temporal cortex, fMRI, psychophysics, synchrony

---

Abstract from the 14th International Multisensory Research Forum, The Hebrew University of Jerusalem, Israel, 2013.

[*] Contact author: ryan.andrew.stevenson@gmail.com

 DOI:10.1163/22134808-000S0015

Symposium #4:
Representing the body in the brain

Laurence R. Harris (Chair), Michael Schaefer, Andrea Serino and Regine Zopf

BRILL Multisensory Research 26 Supplement (2013) 28 brill.com/msr

Talk, "Representing the body in the brain" Symposium

Representing the body in the brain: Effects of body distortion on somatosensory processing

Laurence R. Harris [1,*], **Sarah D'Amour** [2], **Lisa Pritchett** [2], **Michael J. Carnevale** [2] **and Vanessa Harrar** [3]

[1] York University, Canada
[2] Psychology, York University, Canada
[3] Psychology, Oxford University, UK

Abstract

How the body is represented in the brain has been an important topic of investigation for over a hundred years, with neurological, philosophical and physiological implications. Tactile processing, at even the lowest level, including localization, detection thresholds and two-point discrimination thresholds, turns out to depend on factors other than the cutaneous receptors. How these properties are affected and by what, can give insight into how the body is represented in the brain. Distortions in localization related to gaze direction suggest spatial coding in multiple reference frames. Distortions in detection and two-point discrimination thresholds related to alteration in perceived body shape indicate distortions in the relative size of different parts of the representation — perhaps related to receptive field changes in a cortical map. I will review several recent studies that together demonstrate that the body's representation in the brain can be matched to various task requirements and that can be dynamically updated to respond to changes in posture and perceived body dimensions.

Keywords

Body representation, somatosensory, tactile, hand, proprioception, tool use, vision

Abstract from the 14th International Multisensory Research Forum, The Hebrew University of Jerusalem, Israel, 2013.

[*] Contact author: harris@yorku.ca

DOI:10.1163/22134808-000S0016

BRILL Multisensory Research 26 Supplement (2013) 29 brill.com/msr

Talk, "Representing the body in the brain" Symposium

Interaction of embodied cognition and social perception in the primary somatosensory cortex

Michael Schaefer *

University of Magdeburg, Germany

Abstract

In the traditional view our brain represents the body in an area called the primary somatosensory cortex. An increasing body of evidence demonstrated that this brain region represents the perceived body image rather than reflecting mechanical stimulation on the body surface. It has been shown that the primary somatosensory cortex plays a role in multisensory integration of visuotactile stimuli, in empathy, and even in social perception. Here we want to further examine the role of the somatosensory cortices in social perception. We present results of an fMRI study showing how incidental touch sensations influence social judgments in a metaphor-specific way. Participants experienced rough or smooth touch before being asked to judge an ambiguous social interaction. Results revealed that rough touch made social interactions appear more difficult and adversarial, consistent with a rough metaphor (compared with smooth touch or no touch). This impact of tactile cues on social impressions was accompanied by a network including primary somatosensory cortex, amygdala, hippocampus and right inferior prefrontal cortex. Hence, receiving touch from a rough object seems to trigger the application of associated ontological concepts (or scaffolds) even for unrelated people and situations (but not to unrelated or more general feelings). Since this priming was based on somatosensory brain areas, our results provide support for the theory that sensorimotor grounding is intrinsic to cognitive processes.

Keywords

Body, somatosensory cortex, social perception

Abstract from the 14th International Multisensory Research Forum, The Hebrew University of Jerusalem, Israel, 2013.

* Contact author: mischa@neuro2.med.uni-magdeburg.de

DOI:10.1163/22134808-000S0017

BRILL Multisensory Research 26 Supplement (2013) 30 brill.com/msr

Talk, "Representing the body in the brain" Symposium

Multisensory representation of the body in space

Andrea Serino *

Center for Neuroprosthetics, Ecole Polytechnique Fédérale de Lausanne & Department of Psychology, University of Bologna, Switzerland

---

Abstract

The space immediately surrounding the body, i.e., peripersonal space (PPS), is represented by multisensory neural mechanisms integrating tactile, auditory and visual stimuli, presented on or close to the body. I will show that these systems in humans are implemented in a (multi)sensory-motor network of brain areas involving the premotor and posterior-parietal cortices. I will show that PPS representation is plastic, as its boundaries adapt as a function of experience: they extend after that a tool is used to act upon the far space, they contract in case of amputation, and further accommodate after prosthesis implantation. Finally, I will demonstrate that PPS boundaries are sensitive to the presence of and interaction with others.

Keywords

Peripersonal space, body representation

Abstract from the 14th International Multisensory Research Forum, The Hebrew University of Jerusalem, Israel, 2013.

* Contact author: andrea.serino@unibo.it

 DOI:10.1163/22134808-000S0018

BRILL Multisensory Research 26 Supplement (2013) 31 brill.com/msr

Talk, "Representing the body in the brain" Symposium

Representing hands in the brain and effects of viewing hands on temporal synchrony perception

Regine Zopf[1,*], **Jason Friedman**[2] and **Mark A. Williams**[1]

[1] Department of Cognitive Science, Macquarie University, Australia

[2] Department of Physical Therapy, Tel Aviv University, Israel

Abstract

Previous research has shown that specific occipitotemporal areas process visual information of specific body parts such as hands. We used an fMRI block-design to investigate if anterior intraparietal and ventral premotor areas that have been shown to selectively process hand-related synchronous multisensory information also exhibit selective responses to visual hand information. Our findings suggest that both anterior intraparietal and occipitotemporal areas encode visual hand shape and orientation information.

Furthermore, we investigated whether visual hand information modulates temporal synchrony perception. We used a virtual hand setup that includes a glove-based motion capture system (CyberGlove Systems). Participants were asked to perform finger flexion movements and viewed a virtual hand or object performing the same movement. We manipulated the onset delay between the performed and viewed movements and used a two-interval forced choice task to study delay detection performance. Our results indicate that visual spatial hand information modulates the perceptual sensitivity of temporal relationships between self-movements and viewed movements.

Keywords

Body representation, somatosensory, tactile, proprioception, vision

Abstract from the 14th International Multisensory Research Forum, The Hebrew University of Jerusalem, Israel, 2013.

* Contact author: regine.zopf@mq.edu.au

 DOI:10.1163/22134808-000S0019

Symposium #5:
Crossmodal plasticity following sensory loss and restoration

Stephen Lomber (Chair), Olivier Collignon, Andrej Kral and Shelly Levy-Tzedek

BRILL Multisensory Research 26 Supplement (2013) 34 brill.com/msr

Talk, "Crossmodal plasticity following sensory loss and restoration" Symposium

Crossmodal cortical plasticity in the developing brain following sensory loss

Stephen G. Lomber *

Brain and Mind Institute, University of Western Ontario, London, Canada

Abstract

Studies of deaf or blind subjects often report enhanced perceptual abilities in the remaining senses. Compared to hearing subjects, psychophysical studies have revealed specific superior visual abilities in the early-deaf as well as enhanced auditory functions in the early-blind. The neural substrate for these superior sensory abilities has been hypothesized to reside in the deprived cerebral cortices that have been reorganized by the remaining sensory modalities through crossmodal plasticity. In this context, it has been proposed that auditory cortex of the deaf may be recruited to perform visual functions. However, a causal link between supranormal visual performance and the visual activity in the reorganized auditory cortex has never been demonstrated. Furthermore, if auditory cortex does mediate the enhanced visual abilities of the deaf, it is unknown if these functions are distributed uniformly across deaf auditory cortex, or if specific functions can be differentially localized to distinct portions of the affected cortices. These fundamental questions are of significant clinical importance now that restoration of hearing in prelingual deaf children is possible through cochlear prosthetics. Psychophysical, neuroanatomical, electrophysiological, and functional imaging studies will be described that demonstrate crossmodal plasticity in auditory cortex underlies the enhanced visual abilities of the early deaf.

Keywords

Auditory cortex, hearing, deafness

Abstract from the 14th International Multisensory Research Forum, The Hebrew University of Jerusalem, Israel, 2013.

* Contact author: steve.lomber@uwo.ca

DOI:10.1163/22134808-000S0020

BRILL Multisensory Research 26 Supplement (2013) 35–36 brill.com/msr

Talk, "Crossmodal plasticity following sensory loss and restoration" Symposium

Impact of visual experience on the functional organization and the connectivity of the occipital cortex

Olivier Collignon *

Center for Mind/Brain Sciences (CIMeC), University of Trento, Italy

Abstract

How specific brain regions attain, maintain and modify their functional tuning regarding the processing of specific stimuli has long fascinated researchers in neurosciences. Contrasting the impact of congenitally *versus* lately acquired blindness provides a unique model to probe how experience at different developmental periods shapes the functional organization of the occipital cortex (classically considered as visual). I will demonstrate that visual deprivation, acquired early or late in life, induces massive recruitment of occipital regions for auditory processing. Importantly, right dorsal occipital regions, which are known to be preferentially involved in the processing of the spatial relations among visual objects in sighted individuals, are preferentially involved during tasks involving the spatial processing of non-visual inputs in congenitally blind individuals. However such functional specialization in reorganized occipital regions was not observed in the late blind, suggesting that vision has to be lost during an early sensitive period in order to transfer its functional specialization for space processing toward a non-visual modality. By using a combination of functional (psychophysiological interactions) and effective (dynamic causal modeling with Bayesian model selection) connectivity techniques, we also demonstrate that the developmental period of visual deprivation not only influence the functional architecture, but also the connectivity of the occipital cortex. I will finally illustrate how such findings are clinically important now that a growing number of medical interventions may restore vision after a period of visual deprivation.

Keywords

Blind, crossmodal plasticity, occipital, connectivity, auditory

---

Abstract from the 14th International Multisensory Research Forum, The Hebrew University of Jerusalem, Israel, 2013.

* Contact author: oli.collignon@gmail.com

DOI:10.1163/22134808-000S0021

References

Collignon, O., Lassonde, M., Lepore, F., Bastien, D. and Veraart, C. (2007). Functional cerebral reorganization for auditory spatial processing and auditory substitution of vision in early blind subjects, *Cereb. Cortex* **17**(2), 457–465.

Collignon, O., Vandewalle, G., Voss, P., Albouy, G., Charbonneau, G., Lassonde, M., *et al.* (2011). Functional specialization for auditory-spatial processing in the occipital cortex of congenitally blind humans, *Proc. Natl. Acad. Sci. USA* **108**(11), 4435–4440.

Dormal, G. and Collignon, O. (2011). Functional selectivity in sensory-deprived cortices, *Journal of Neurophysiology* **105**(6), 2627–2630.

BRILL Multisensory Research 26 Supplement (2013) 37 brill.com/msr

Talk, "Crossmodal plasticity following sensory loss and restoration" Symposium

Cross-modal plasticity in the congenitally deaf cat

A. Kral [1,*], **R. Land** [1], **P. Baumhoff** [1], **J. Tillein** [1], **P. Hubka** [1] **and S. G. Lomber** [2]

[1] Institute of Audioneurotechnology & Dept. of Experimental Otology, ENT Clinics, Hannover School of Medicine, Hannover, Germany

[2] Department of Physiology and Pharmacology, School of Medicine, University of Western Ontario, Canada

Abstract

The experiments were undertaken on 8 cats (4 deaf, 4 hearing), recordings were from dorsal auditory cortex (field DZ) and an adjoining visual field (PMLS) under cochlear implant (auditory) and visual stimulation in light isoflurane/N_2O anesthesia (Land *et al.*, 2012, *PLoS One*). Visual stimuli were flashes (increasing and decreasing luminance) phase-reversal gratings with different orientations and spatial frequencies. Simultaneous recordings were taken using two 16-channel Neuronexus probes inserted at ~10 positions in both fields spatially as close as possible. The DZ penetration had an orientation parallel to the microcolumns. Electrodes were stained with DiI or DiO and the penetrations were histologically reconstructed after the experiment. Recording sites within DZ were confirmed using an SMI-32 staining.

From all recording sites, only the sites containing unit activity were processed further. Altogether, 1424 unit responses were evaluated, ~700 in each group. Spontaneous activity was significantly higher in PMLS than in DZ in both groups of animals. Unit responses showing statistically significant correlation with current level were considered responsive to auditory stimulation. There were significant auditory responses in both groups of animals in DZ (deaf: 29%; hearing: 36% of sites). In PMLS, only few units responded to auditory stimulation (deaf: 2%; hearing: 4%). Visual stimulation led to responses in one third of units in PMLS in both groups of animals (deaf: 31%; hearing: 29%). In DZ, on the other hand, the deaf cats had more visual responses (deaf: 7%; hearing: 2%). In these responsive units, the evoked firing rate was higher in deaf cats, whereas hearing cats demonstrated more a modulation of ongoing activity than an evoked response in DZ. Only few bimodal units responsive to both auditory and visual stimulation were observed in DZ of deaf cats.

The present results suggest a modest visual cross-modal reorganization of field DZ in congenital deafness. However, the auditory responsiveness was preserved in this field, too. The relatively small number of visually-responsive cells, corresponding to previous tracer studies (Barone *et al.*, 2013, *PLoS One*), demonstrates that also other (possibly attentional) processes are involved in supranormal visual performance of deaf subjects.

Abstract from the 14th International Multisensory Research Forum, The Hebrew University of Jerusalem, Israel, 2013.

* Contact author: akral@mac.com

 DOI:10.1163/22134808-000S0022

BRILL Multisensory Research 26 Supplement (2013) 38 brill.com/msr

Talk, "Crossmodal plasticity following sensory loss and restoration" Symposium

The action-perception loop examined across senses

Shelly Levy-Tzedek [1,2,*] **and Amir Amedi** [1,2,3]

[1] The Edmond and Lily Safra Center for Brain Sciences (ELSC), The Hebrew University of Jerusalem, Jerusalem, Israel

[2] Department of Medical Neurobiology, The Institute for Medical Research Israel–Canada (IMRIC), Faculty of Medicine, The Hebrew University of Jerusalem, Jerusalem, Israel

[3] The Cognitive Science Program, The Hebrew University of Jerusalem, Jerusalem, Israel

---

Abstract

We examined how information flowing through the action-perception loop transfers across senses. I will review the results of our studies exploring movement guided by vision compared to movement guided by the EyeMusic sensory-substitution device, and how the two may interact. I will argue that sensory-motor learning is not sensory-modality-specific, and that novel sensory-motor information can be transferred between sensory modalities. I will conclude by outlining the important implications of these findings for visual rehabilitation, and for the visually impaired.

Keywords

SSD, sensory substitution, motor control, action-perception, visuomotor transfer, sensory restoration

References

Levy-Tzedek, S., Novick, I., Arbel, R., Abboud, S., Maidenbaum, S., Vaadia, E. and Amedi, A. (2012). Cross-sensory transfer of sensory-motor information: visuomotor learning affects performance on an audiomotor task, using sensory-substitution, *Scientific Reports* **2**, 949.

Levy-Tzedek, S., Hanassy, S., Abboud, S., Maidenbaum, S. and Amedi, A. (2012). Fast, accurate reaching movements with a visual-to-auditory sensory substitution device, *Restorative Neurology and Neuroscience* **30**(4), 313–323.

Abstract from the 14th International Multisensory Research Forum, The Hebrew University of Jerusalem, Israel, 2013.

[*] Contact author: shelly.le@mail.huji.ac.il

DOI:10.1163/22134808-000S0023

Symposium #6: Hyper-binding across the senses

Fiona Newell, Romke Rouw (co-Chairs), Devin Terhune and Olympia Colizoli

BRILL Multisensory Research 26 Supplement (2013) 40 brill.com/msr

Talk, "Hyper-binding across the senses" Symposium

Multisensory interactions and synesthesia

Fiona Newell * and David McGovern

School of Psychology and Institute of Neuroscience, Trinity College Dublin, Ireland

Abstract

Although the behavioural and neurological processes underpinning the condition of synaesthesia have been elucidated in recent years, there have been relatively fewer attempts to investigate the extent to which synaesthesia is mediated by multisensory processes common to all. Neural models propose connectivity patterns that may be specific to the condition, yet some recent behavioural evidence suggests that synaesthetic interactions may be scaffolded by general perceptual mechanisms. Here I will discuss some of our recent evidence supporting common genetic links between different types of synaesthesia which suggests that different synaesthetic experiences may be based on the same ontogeny. Moreover, evidence for differences in early perceptual processing between synaesthetes and non-synaesthetes suggests that synaesthesia is not as discrete a condition as previously thought. Our findings further suggest that synaesthesia is induced by multisensory rather than unisensory inputs, and that the synaesthetic experience may be penetrated by ongoing veridical perceptual processes. Other evidence for shared synaesthetic-like associations across the senses is consistent with the idea that synaesthesia is not a discrete condition, but may depend on interactions across the senses common to all, the extent to which is not yet fully understood.

Abstract from the 14th International Multisensory Research Forum, The Hebrew University of Jerusalem, Israel, 2013.

* Contact author: fnewell@tcd.ie

 DOI:10.1163/22134808-000S0024

BRILL Multisensory Research 26 Supplement (2013) 41 brill.com/msr

Talk, "Hyper-binding across the senses" Symposium

Cortical excitability, neurochemistry, and enhanced cognition: A lesson from synesthesia

Roi Cohen-Kadosh * **and Devin Terhune**

Department of Experimental Psychology, University of Oxford, UK

Abstract

Primary visual cortex has been repeatedly implicated in grapheme-colour synaesthesia but the role it plays in supporting different characteristics of this condition, and its connection with perceptual awareness, remain poorly understood. Here I will describe recent research concerning cortical excitability and neurochemical concentrations in primary visual cortex and their cognitive and phenomenological correlates in synaesthesia. I will further demonstrate that modulating cortical excitability in primary visual cortex modifies the synaesthetic experience. This research sheds light on the role of primary visual cortex in synaesthesia and its link to consciousness, and provides further information regarding the neural basis of individual differences and enhanced cognition in this condition.

Abstract from the 14th International Multisensory Research Forum, The Hebrew University of Jerusalem, Israel, 2013.

* Contact author: roi.cohenkadosh@psy.ox.ac.uk

 DOI:10.1163/22134808-000S0025

BRILL Multisensory Research 26 Supplement (2013) 42 brill.com/msr

Talk, "Hyper-binding across the senses" Symposium

Pseudo-synesthesia through reading books with colored letters and experience-dependent plasticity of the visual system

Olympia Colizoli *, **Jaap M. J. Murre**, **H. Steven Scholte and Romke Rouw**
Department of Psychology, University of Amsterdam, The Netherlands

Abstract
Grapheme–color synesthesia is defined as the experience of color in relation to letters, words and numbers. This form of synesthesia is a relatively common type and has received the most attention in the research field in terms of the neural basis of the synesthetic experience and behavioral features of synesthesia such as attention and memory. Grapheme–color and linguistic–color synesthesia have a genetic component. Still, language (the most common synesthetic inducer) is acquired through an interaction between genes and the environment. In order to probe this interaction, we compared relatives of grapheme–color synesthetes to a group of matched controls in a reading-in-color training paradigm. All participants read specially prepared books in which four high-frequency letters were paired with four high-frequency colors. Magnetic resonance images (MRI) were acquired before and after training, including diffusion tensor imaging (DTI). In this sample, there was no evidence of increased acquisition of the letter–color associations by the relatives of synesthetes. No group differences were found in behavior, while there are differences in functional activation on the Stroop task. Imaging results showed significant differences in brain activation in regions known to be involved in grapheme–color synesthesia across both groups. These regions include temporal–occipital cortex, inferior and superior parietal lobe, precentral gyrus, and frontal pole. We conclude that the neural basis of the effects of reading in color relies primarily on the same brain regions as the experience of (authentic) grapheme–color synesthesia.

Keywords
Synesthesia, grapheme, color, reading, training, learning, memory, visual

Abstract from the 14th International Multisensory Research Forum, The Hebrew University of Jerusalem, Israel, 2013.

* Contact author: olympia.colizoli@gmail.com

 DOI:10.1163/22134808-000S0026

BRILL Multisensory Research 26 Supplement (2013) 43 brill.com/msr

Talk, "Hyper-binding across the senses" Symposium

The neural basis of different types of synesthesia

Romke Rouw * **and Steven Scholte**

Department of Psychology, University of Amsterdam, The Netherlands

Abstract

The condition synesthesia offers an extraordinary opportunity to study mechanisms involved in (hyper)-binding across different modalities. Currently, the majority of synesthesia research is aimed at a few types of synesthesia; colors evoked by written or spoken language and spatial arrangements evoked by sequential concepts. We do not yet know whether the processes involved in multisensory binding are similar or different across different types of synesthesia. In this project, brain structure measurements (connectivity and volume) were obtained from a large group of subjects (over 350 individuals). Furthermore, the subjects were presented with an extensive synesthesia questionnaire. In this subject group, different types of synesthetes were present; the inducers and concurrent ranged across perceptual modalities (e.g., visual, auditory, taste, tactile) but also involved non-sensory concepts (e.g., 'personality'). This allows studying which of the previously obtained white-matter and grey-matter differences in synesthesia are specific to that (i.e., linguistic–color) type of synesthesia, and which are related more generally to having synesthesia. Furthermore, it allows examining whether synesthetes are, as a group, categorically different from all non-synesthetic subjects. Alternatively, each synesthete is only different from a non-synesthete in the binding of his or her specific inducer to his or her specific concurrent.

Abstract from the 14th International Multisensory Research Forum, The Hebrew University of Jerusalem, Israel, 2013.

* Contact author: R.Rouw@uva.nl

 DOI:10.1163/22134808-000S0027

Symposium #7: Temporal recalibration to asynchronous multisensory stimuli

Erik van der Burg (Chair), John Cass, Marc Ernst, Carmel Levitan and Jean Vroomen

BRILL Multisensory Research 26 Supplement (2013) 46 brill.com/msr

Talk, "Temporal recalibration to asynchronous multisensory stimuli" Symposium

Adaptation to audiovisual asynchrony occurs rapidly

Erik van der Burg [1,*], **David Alais** [1] **and John Cass** [2]

[1] School of Psychology, University of Sydney, Australia

[2] School of Psychology, University of Western Sydney, Australia

---

Abstract

In order to combine information from different sensory modalities the brain must deal with considerable temporal uncertainty. In natural environments, an external event may produce simultaneous auditory and visual signals yet they will invariably activate the brain asynchronously due to different propagation speeds for light and sound and different neural response latencies once the signals reach the observer. One strategy the brain uses to deal with audiovisual timing variation is to adapt to a prevailing asynchrony to help re-align the signals. Here we investigate audiovisual recalibration and show that it takes place extremely rapidly. Our results demonstrate that exposure to a single, brief asynchrony is sufficient to produce strong recalibration effects. Such fast-acting recalibration provides a mechanism for overcoming inevitable audiovisual timing variation and serves to rapidly re-align signals at onset to maximize the perceptual benefits of audiovisual integration.

Keywords

Temporal recalibration, audition, vision

References

Fujisaki, W., Shimojo, S., Kashino, M. and Nishida, S. (2004). Recalibration of audiovisual simultaneity, *Nature Neuroscience* **7**(7), 773–778.

Vroomen, J., Keetels, M., De Gelder, B. and Bertelson, P. (2004). Recalibration of temporal order perception by exposure to audio-visual asynchrony, *Cognitive Brain Research* **22**, 32–35.

Abstract from the 14th International Multisensory Research Forum, The Hebrew University of Jerusalem, Israel, 2013.

* Contact author: erik.vanderburg@sydney.edu.au

 DOI:10.1163/22134808-000S0028

BRILL

Multisensory Research 26 Supplement (2013) 47

MULTISENSORY RESEARCH

brill.com/msr

Talk, "Temporal recalibration to asynchronous multisensory stimuli" Symposium

Audio-visual temporal recalibration in cluttered displays

John Cass [1,*], **Diane Oake** [1] **and Erik van der Burg** [2]

[1] School of Psychology, University of Western Sydney, Australia

[2] School of Psychology, University of Sydney, Australia

Abstract

Previous demonstrations of temporal recalibration have used single audiovisual events. Natural environments, however, are often highly cluttered both spatially and temporally. We investigated recalibration in highly cluttered visual scenes, examining to what extent it is retinotopically specific. Each trial consisted of an adaptation phase, where participants were exposed to 0.7 Hz square-wave amplitude modulating 500 Hz pure tone and luminance-defined disk (0.7 Hz) for 60 s; and a test phase. We manipulated the relative phase of the adapting tone and disk, corresponding to A-V SOAs of −106, 0 and +106 ms, and the location of the adapting disk stimulus (at fixation *vs.* 14° to the right of fixation). The test stimulus consisted of 20 luminance-defined disks concentrically arranged around fixation (14° radius). The tone and luminance of each disk modulated abruptly (0.7 Hz). Each disk was presented at a unique temporal phase with each temporally adjacent pair separated by 40 ms. On each trial, the phase of one disk matched that of the tone modulation. Critically, the temporal phase associated with each disk location varied randomly from trial-to-trial and it was the subjects' task to identify the synchronized disk. We measured the proportion of trials for which a given disk was chosen as being synchronous. Shifts in PSS of ∼−50 ms were observed following tone-first adaptation and ∼+120 ms following flash-first adaptation. Visual quadrant analysis revealed no differences in the PSS shifts following foveal and peripheral adaptation. We conclude that recalibration may be amplified by the presence of spatio-temporal visual clutter, and this is unconstrained by the location of adapting stimuli.

Keywords

Audio-visual, temporal recalibration, psychophysics

Abstract from the 14th International Multisensory Research Forum, The Hebrew University of Jerusalem, Israel, 2013.

* Contact author: johncassvision@gmail.com

DOI:10.1163/22134808-000S0029

BRILL Multisensory Research 26 Supplement (2013) 48 brill.com/msr

Talk, "Temporal recalibration to asynchronous multisensory stimuli" Symposium

Does causal binding affect temporal recalibration?

Marc Ernst * **and Marieke Rohde**

Bielefeld University, Germany

Abstract

Temporal recalibration has been shown for multisensory stimuli that are passively perceived such as visual–haptic or visual–auditory events. We have recently proposed models based on Bayesian inference showing that these recalibration mechanisms are in line with optimal estimation schemes. Similarly temporal recalibration has been shown for sensorimotor events. Sensorimotor events are unique in the way that when acting on the world we are the cause for the changes that occur and that we perceive. There is a unique temporal order between the cause–effect relationship, which is between the action and subsequent percept of the change. In some recent studies we investigated how this cause–effect relationship impacts on the temporal recalibration of sensorimotor recalibration.

Abstract from the 14th International Multisensory Research Forum, The Hebrew University of Jerusalem, Israel, 2013.

* Contact author: marc.ernst@uni-bielefeld.de

 DOI:10.1163/22134808-000S0030

BRILL

Multisensory Research 26 Supplement (2013) 49–50

MULTISENSORY RESEARCH

brill.com/msr

Talk, "Temporal recalibration to asynchronous multisensory stimuli" Symposium

Cross-modal temporal frequency channels for rate classification

Carmel A. Levitan [1,*], Charlotte L. Yang [2], Yih-Hsin Alison Ban [1], Noelle R. B. Stiles [2] and Shinsuke Shimojo [2]

[1] Cognitive Science Program, Occidental College, USA

[2] Computation and Neural Systems Program, California Institute of Technology, USA

Abstract

We previously reported our discovery that temporal rate adaptation transfers bidirectionally between vision and audition. Temporal frequency channels are linked across audition and vision (Yao *et al.*, 2009); but duration channels for audition and vision are thought to be independent (Heron *et al.*, 2012). We used our paradigm to characterize linkages between auditory and visual channels by measuring whether or not transfer of adaptation still occurs as the discrepancy between adaptation and test frequencies increases. Participants ran in three experimental sessions, each with a different adaptation frequency. They were trained, using feedback, to classify flickering visual stimuli (ranging in frequency from 3.25–4.75 Hz) as fast or slow (relative to 4 Hz). They then classified 140 pre-adaptation test trials with feedback, providing a baseline. Afterwards, 30 adaptation trials of auditory stimuli beeping at either 5, 8, or 12 Hz were presented, followed by 20 alternating blocks of 7 adaptation and 7 post-adaptation test trials (without feedback). We compared the PSE of the pre- and post-adaptation trials to quantify the cross-modal transfer and found that the aftereffect occurred when the adaptation frequency was most similar to the test frequencies but was no longer present with larger discrepancies. These results rule out response bias as a plausible explanation for our original findings and suggest that the timing mechanisms underlying rate perception are consistent with supramodal channels that are tuned.

Keywords

Temporal processing, rate adaptation, channels, adaptation, cross-modal transfer

Abstract from the 14th International Multisensory Research Forum, The Hebrew University of Jerusalem, Israel, 2013.

* Contact author: levitan@gmail.com

DOI:10.1163/22134808-000S0031

References

Heron, J., Aaen-Stockdale, C., Hotchkiss, J., Roach, N. W., McGraw, P. V. and Whitaker, D. (2012). Duration channels mediate human time perception, *Proceedings of the Royal Society B* **279**, 690–698.

Yau, J. M., Olenczak, J. B., Dammann, J. F. and Bensmaia, S. J. (2009). Temporal frequency channels are linked across vision and touch, *Current Biology* **19**(7), 561–566.

BRILL Multisensory Research 26 Supplement (2013) 51 brill.com/msr

Talk, "Temporal recalibration to asynchronous multisensory stimuli" Symposium

Explicit judgments and implicit measures of temporal recalibration

Jean Vroomen * **and Mirjam Keetels**

Tilburg University, The Netherlands

Abstract

The timing relation between sensory modalities, including motor actions and the sensory consequences of that action, can be adapted by exposing participants to artificial delays in the involved modalities (i.e., temporal recalibration). The mechanism underlying temporal recalibration, though, remains elusive. In our research, we examine to which extent explicit tests about temporal relation — like temporal order or synchrony judgments — yield similar result as more implicit test. We address the intriguing question whether explicit judgments about crossmodal temporal order *versus* implicit measures can actually dissociate.

Abstract from the 14th International Multisensory Research Forum, The Hebrew University of Jerusalem, Israel, 2013.

* Contact author: J.Vroomen@uvt.nl

DOI:10.1163/22134808-000S0032

Symposium #8:
Development and plasticity in multisensory processes

Mark Wallace (Chair), Andy Bremner, David J. Lewkowicz,
Charles Spence and Terrence R. Stanford

BRILL Multisensory Research 26 Supplement (2013) 54 brill.com/msr

Talk, "Development and plasticity in multisensory processes" Symposium

Development of multisensory temporal processing: Bridging between animal and human studies

Mark T. Wallace [1,2,3,4,*], **Andrea Hillock-Dunn** [5], **Dipanwita Ghose** [2], **Juliane Krueger-Fister** [4], **Diana Sarko** [1,4] **and Ryan Stevenson** [1,4]

[1] Department of Hearing & Speech Sciences, Vanderbilt University, Nashville, TN, USA
[2] Department of Psychology, Vanderbilt University, Nashville, TN, USA
[3] Department of Psychiatry, Vanderbilt University, Nashville, TN, USA
[4] Vanderbilt Brain Institute, Vanderbilt University, Nashville, TN, USA
[5] Department of Speech & Hearing Sciences, University of North Carolina, Chapel Hill, NC, USA

Abstract

The talk will compare and contrast results from animal neurophysiological and human psychophysical studies that have focused on characterizing the maturation of temporal factors that are critical determinants in the integration of audiovisual stimuli. In the animal model realm, individual neurons in both cortical and subcortical structures progress through a series of developmental stages. During this maturational progression, neurons first transition from having temporal tuning profiles that are quite narrow to having tuning functions that are very broad. This initial period of broadening is then followed by a more protracted developmental period in which these temporal tuning profiles gradually narrow to achieve adult-like values. In human development, where we have studied children from ages 6–18, a gradual narrowing of multisensory temporal function, specifically in the width of the multisensory temporal binding window, is observed. Our evidence suggests that this window is not fully mature until late adolescence, highlighting a surprisingly long period during which multisensory temporal function is changing. These results will be discussed in the context of their implications for both behavioral and perceptual development.

Keywords

Maturation, audiovisual, temporal binding, neurophysiology, behavior, perception

Abstract from the 14th International Multisensory Research Forum, The Hebrew University of Jerusalem, Israel, 2013.

* Contact author: mark.wallace@vanderbilt.edu

 DOI:10.1163/22134808-000S0033

BRILL Multisensory Research 26 Supplement (2013) 55 brill.com/msr

Talk, "Development and plasticity in multisensory processes" Symposium

Developmental change in multisensory body representations in early childhood

Andrew J. Bremner [1,*], **Jannath Begum** [1], **Charles Spence** [2] **and Dorothy Cowie** [1]

[1] Department of Psychology, Goldsmiths, University of London, UK
[2] Department of Experimental Psychology, University of Oxford, UK

Abstract
A number of recent findings indicate that the multisensory processes underlying our ability to localize our limbs continue to undergo development beyond infancy and across childhood. Here we review findings from our recent investigations of the development of multisensory bodily illusions in early life (see Bremner *et al.*, 2013; Cowie *et al.*, in press). Studies of the mirror hand illusion and the rubber hand illusion in children show that children's reliance on visual information about the hand continues to change dramatically until 12 to 13 years of age. There are early increases up to 5 years in children's reliance on visual cues (in the mirror hand illusion), followed by a much greater reliance on visual cues (in the rubber hand illusion even when the fake hand is presented in an implausible posture) in middle childhood, and a subsequent decline into early adolescence and adulthood. In explaining these developmental changes we discuss the development of representations of the canonical layout of the body, and an ability to represent the limbs across dynamic changes in posture, making specific reference to the changing role of vision in guiding action in the environment across early childhood.

Keywords
Development, spatial representation, body representation

References

Bremner, A. J., Hill, E. L., Pratt, E., Rigato, S. and Spence, C. (2013). Bodily illusions in young children: Developmental change in the contribution of vision to perceived hand position in early childhood, *PLoS ONE* **8**, e51887. DOI:10.1371/journal.pone.0051887

Cowie, D., Makin, T. and Bremner, A. J. (in press). Children's responses to the Rubber Hand Illusion reveal dissociable pathways in body representations, *Psychological Science*.

Abstract from the 14th International Multisensory Research Forum, The Hebrew University of Jerusalem, Israel, 2013.

* Contact author: a.bremner@gold.ac.uk

 DOI:10.1163/22134808-000S0034

BRILL Multisensory Research 26 Supplement (2013) 56 brill.com/msr

Talk, "Development and plasticity in multisensory processes" Symposium

How self-generated experience contributes to the development of speech and language in human infants?

David J. Lewkowicz [1,2,*]

[1] Department of Psychology, Florida Atlantic University, Florida, USA
[2] Center for Complex Systems & Brain Sciences, Florida Atlantic University, Florida, USA

Abstract
Everyday experience plays a central role in the development of perceptual and cognitive abilities. Our latest studies have been exploring the non-obvious and obvious effects of perceptual experience on the development of audiovisual speech and language perception in human infants. Using eye tracking technology, we have discovered that when infants begin babbling and, thus, start learning to speak their native language, they begin to shift their attention to the interlocutor's mouth to access redundant, audiovisual speech cues. This shift is self-generated and, thus, is evidence of the non-obvious effects of experience. We have also found that once infants start to master their native speech forms and begin to acquire native-language expertise, they begin to shift their attention to the interlocutor's eyes to access social cues. They only make this shift, however, in response to their native language reflecting the specific effects of native-language experience. In this talk, I will present the latest findings from several new studies in which we have explored bilingual infants' shifting patterns of audiovisual attention, the separate contribution of the audible and visible attributes of audiovisual speech information to the developmental shift in selective attention, and the importance of the temporal synchrony of the audible & visible attributes of audiovisual speech to the observed changing developmental pattern of selective attention. Together, these findings provide new insights into the mechanisms underlying the developmental acquisition of audiovisual speech & language skills in human infancy.

Keywords
Development, audiovisual perception, human infants, speech, language

Abstract from the 14th International Multisensory Research Forum, The Hebrew University of Jerusalem, Israel, 2013.
* Contact author: lewkowic@fau.edu

 DOI:10.1163/22134808-000S0035

BRILL Multisensory Research 26 Supplement (2013) 57 brill.com/msr

Talk, "Development and plasticity in multisensory processes" Symposium

Questioning the utility of the concept of amodality: Towards a revised framework for understanding crossmodal relations

Charles Spence [1,*], **Ophelia Deroy** [2] **and Andy Bremner** [1]

[1] Department of Experimental Psychology, University of Oxford, UK

[2] Centre for the Study of the Senses, Institute of Philosophy, London, UK

Abstract

The idea that sensory experiences should be distinguished into modal and amodal (in terms of crossmodal redundancy, rather than amodal completion) runs through many areas of psychology; developmental psychologists, in particular, appear to hold opposing views about whether infants start by having modally distinct or amodal experiences of the properties that are commonly considered to be amodal, such as space, duration and intensity. Many other phenomena in adults — such as 'cross-modal correspondences' (Spence, 2011) — are deemed interesting, in part, precisely because they are not simply picking-up on amodal properties or redundantly-coded features. However, while many researchers have, over the years, used the term 'amodal', we would argue that there is a great deal of uncertainty over which dimensions of experience should be considered as amodal, and according to which criteria. Here, we argue that virtually nothing is amodal. To the extent that this claim holds, it may help to shed further light on early multisensory development while at the same time calling into question various theoretical accounts about both the acquisition of crossmodal relations in early development and the acquisition of novel associations (or correspondences) in adulthood. This discussion may also help to elucidate the kinds of evidence that can legitimately be taken to support arguments concerning the existence, or otherwise, of neonatal synaesthesia (see Deroy and Spence, submitted).

Keywords

Development, amodal, perception, crossmodal correspondences, synaesthesia

References

Deroy, O. and Spence, C. (submitted). From buzzing confusion to synaesthetic confusion: Examining the neonatal synaesthesia hypothesism, *Neuroscience & Biobehavioral Reviews*.

Spence, C. (2011). Crossmodal correspondences: A tutorial review, *Attention, Perception, & Psychophysics* **73**, 971–995.

Abstract from the 14th International Multisensory Research Forum, The Hebrew University of Jerusalem, Israel, 2013.

* Contact author: charles.spence@psy.ox.ac.uk

 DOI:10.1163/22134808-000S0036

BRILL Multisensory Research 26 Supplement (2013) 58 brill.com/msr

Talk, "Development and plasticity in multisensory processes" Symposium

Cross-modal covariance as a key for the maturation of multisensory integration

Terrence R. Stanford *, **Xinghong Xu**, **Benjamin Rowland** and **Barry E. Stein**

Dept. of Neurobiology & Anatomy, Wake Forest School of Medicine,
Winston-Salem, NC, USA

Abstract

The ability to integrate information derived from different senses matures over a protracted period of postnatal development. In the superior colliculus (SC), a midbrain structure responsible detecting and orienting to sensory events, this maturational profile is characterized by the gradual appearance of neurons sensitive to more than one modality, and later, by the ability of these neurons to integrate the modality-specific information they receive. It is likely that associative learning principles are at play in guiding multisensory development as the organism gains experience with cross-modal cues. To gain insight into the critical experiential requirements, cats were reared in omnidirectional sound (i.e., noise), a sensory landscape designed to provide co-activation of visual (normal, and discrete) and auditory (ambient, and diffuse) inputs to SC neurons, but to obviate a key requirement of Hebbian-based learning rules *via* the imposition of uncorrelated noise on one sensory-specific channel (i.e., auditory). The present findings establish that co-activation of sensory-specific SC input channels is not sufficient for SC neurons to develop the ability to integrate cross-modal information. By decoupling cross-modal co-activation and cross-modal co-variance, the noise rearing paradigm provides a key insight lacking from prior studies using dark-rearing or random exposure to visual and auditory stimuli. Thus, the data provide strong evidence that the maturation of SC multisensory integration capabilities requires the encoding of statistical regularities between the cross-modal cues that characterize the rearing environment.

Abstract from the 14th International Multisensory Research Forum, The Hebrew University of Jerusalem, Israel, 2013.

* Contact author: stanford@wakehealth.edu

DOI:10.1163/22134808-000S0037

Section II: ORAL PRESENTATIONS

BRILL Multisensory Research 26 Supplement (2013) 60 brill.com/msr

Oral Presentation

Temporal recalibration involves adaptation at two time scales

David Alais [1,*], **John Cass** [2] **and Erik van der Burg** [1]

[1] University of Sydney, Australia
[2] University of Western Sydney, Australia

Abstract

We investigate the time constant of recovery from adaptation to temporal asynchrony. Subjects adapted to a 4-min movie of a nail being hammered, a naturalistic animation with strong audiovisual temporal cues. The movie soundtrack had a constant asynchrony of either ±200 ms. Following adaptation, we sampled synchrony judgments every 2 seconds for 2 min using a brief audiovisual synchrony task (a flash and beep). The flash and beep varied over several ±SOAs and subjects indicated whether they were synchronous. Binning the synchrony responses within a short, rolling time window we obtained estimates of PSS over the course of recovery from temporal adaptation. The rolling average PSS showed a significant recalibration initially followed by a clear recovery function, with PSSs steadily returning to baseline after ~60 s. We also analysed short-time scale recalibration by testing for adaptation effects between successive synchrony probes. Although these probes were brief (60 ms), we found that a given synchrony judgment in the post adaptation period was strongly influenced by the sign of the previous synchrony probe, showing an adaptation effect in the direction of the preceding synchrony probe's SOA. In sum, these results show long- and short-scale temporal recalibration effects, with the short-scale inter-probe effects superimposed on the long-scale recalibration. In a second experiment, we delayed the synchrony probes until 60 s post-adaptation and observed no long-scale temporal recalibration, showing there is no storage of long-scale temporal adaptation.

Keywords

Recalibration, adaptation, temporal processing

Abstract from the 14th International Multisensory Research Forum, The Hebrew University of Jerusalem, Israel, 2013.

[*] Contact author: davida@psych.usyd.edu.au

 DOI:10.1163/22134808-000S0038

BRILL Multisensory Research 26 Supplement (2013) 61 brill.com/msr

MULTISENSORY RESEARCH

Oral Presentation

Regularization improves models of audiovisual integration in speech perception

Tobias S. Andersen *

Technical University of Denmark, Denmark

Abstract

Visual speech, the speech information mediated by the sight of articulatory mouth movements, can influence the auditory phonetic speech percept. This is demonstrated by the McGurk illusion, where an acoustic speech signal (e.g., /ba/) is perceived differently (as /da/) when perceived audio-visually, dubbed onto the video of an incongruent talking face (articulating /ga/).

A computational account of the integration of information across the senses underlying the McGurk illusion has long been sought. One account, the Fuzzy Logical Model of Perception (FLMP, Massaro, 1998) posits that integration is based on fuzzy truth-values. Here we present an alternative accounts in which integration is based on continuous feature values.

We show that such models can provide a better fit to observed data than the FLMP. In order to take model-flexibility into account, we cross-validate model fits and show that although feature based models have more predictive power than the FLMP, both types of models perform rather poorly. Finally, we show that the predictive power of both types of models improve when the models are regularized by Bayesian priors and that after regularization, feature based models have significantly better predictive power than the FLMP.

Keywords

Audiovisual, speech perception, modeling

Reference

Massaro, D. W. (1998). *Perceiving Talking Faces*. MIT Press, Cambridge, Massachusetts.

Abstract from the 14th International Multisensory Research Forum, The Hebrew University of Jerusalem, Israel, 2013.

* Contact author: toban@imm.dtu.dk

 DOI:10.1163/22134808-000S0039

BRILL Multisensory Research 26 Supplement (2013) 62 brill.com/msr

Oral Presentation

The cost of adopting and adapting tactile frames of reference

Gabriel Arnold [1,*], **Jess Hartcher-O'Brien** [1], **Charles Spence** [2] and **Malika Auvray** [1]

[1] LIMSI – CNRS, France
[2] Oxford University, UK

Abstract

Visual information is predominantly interpreted within an eye-centered reference frame. Tactile information, on the other hand, can be interpreted within different reference frames, i.e., local-surface, or whole-body centered. An important question is whether, given the different possibilities, each observer has a natural reference frame that they consistently adopt in different conditions, or they can freely adopt several reference frames and switch from one to another without cost. Recognition of ambiguous asymmetrical tactile letters (e.g., b, d, p, q) allows us to interrogate the different reference frames adopted by observers when interpreting tactile information. For such stimuli drawn on the skin, recognition requires assigning top–bottom, left–right and front–back axes to the letter. Across several experiments, participants had to recognize these letters when presented on different body surfaces, either with a freely adopted reference frame or an imposed one. In the unconstrained condition, participants consistently adopted one reference frame, with between-participant differences being evident. When adapted to a new reference frame, interpretation of the letters produced an important cost in response accuracies and latencies, indicating that the freely adopted reference frame corresponded to a natural reference frame rather than an arbitrary choice. This cost was greater for the freely-adopted egocentric (head- or body-centered) than for the non-egocentric (off-centered) reference frame. By training participants with a particular set of symbols, we then tested for generalization to novel stimuli and body surfaces. Our results have implications for the design of visuo-tactile sensory substitution devices and for understanding emergence of the distal attribution phenomenon.

Keywords

Tactile perception, sensory substitution, learning, generalization, distal attribution

Reference

Parsons, L. M. and Shimojo, S. (1987). Perceived spatial organization of cutaneous patterns on surfaces of the human body in various positions, *Journal of Experimental Psychology: Human Perception and Performance* **13**, 488–504.

Abstract from the 14th International Multisensory Research Forum, The Hebrew University of Jerusalem, Israel, 2013.

* Contact author: gabriel.arnold@cegetel.net

DOI:10.1163/22134808-000S0040

BRILL Multisensory Research 26 Supplement (2013) 63 brill.com/msr

Oral Presentation

Psychophysical evidence for a generalized sense of number

Roberto Arrighi *

University of Florence, Italy

Abstract

Previous studies have shown that numerosity perception is susceptible to adaptation, suggesting that it is a primary visual property (Burr and Ross, 2008). Recent studies on monkeys have revealed a sharp numerosity selectivity for neurons of intraparietal sulcus (IPS), with some tuned to an abstract representation of numerosity independent of the temporal format (simultaneous or sequential) of presentation (Nieder *et al.*, 2006). Here we exploit adaptation technique to test demonstrate a truly abstract number sense in humans. We demonstrate numerosity adaptation for sequentially presented items both across modalities as well as across presentation format. Adapting to a series of flashes presented at 2 or 8 Hz for about 40 seconds changes the perceived numerosity of visual pulses (range 2–20) presented within a temporal window of 2 seconds: adapting to 2 Hz increased perceived number by 15–20%; adapting to 8 Hz decreased perceived number by a similar amount. For adaptation to occur, the adaptor stimulus had to occupy the same spatial position in external rather than retinal coordinates. Similar adaptation effects were observed for auditory stimuli (adaptation to tone sequences), and also cross-modally: adapting to auditory bursts affected perceived visual numerosity and *vice versa*, to similar extent as intra-modal adaption. Finally we demonstrated cross-format adaptation: adapting to sequential streams of flashes affect perceived numerosity of spatial arrays and viceversa. All these results point to the existence of a perceptual system that combines visual and auditory information to encode an abstract sense of number, in space and in time.

Keywords

Numerosity perception, cross-modal perception, adaptation

References

Burr, D. and Ross, J. (2008). A visual sense of number, *Curr. Biol.* **18**(6), 425–428.

Nieder, A., Diester, I. and Tudusciuc, O. (2006). Temporal and spatial enumeration processes in the primate parietal cortex, *Science* **313**(5792), 1431–1435.

Abstract from the 14th International Multisensory Research Forum, The Hebrew University of Jerusalem, Israel, 2013.

* Contact author: roberto.arrighi@gmail.com

DOI:10.1163/22134808-000S0041

BRILL Multisensory Research 26 Supplement (2013) 64 brill.com/msr

MULTISENSORY RESEARCH

Oral Presentation

Tracking the evolution of learning a novel dance sequence in expert ballet dancers

Rachel J. Bar [1,*] **and Joseph F. DeSouza** [2]

[1] Department of Psychology, Ryerson University, Toronto, ON, Canada
[2] Centre for Vision Research, York University, Canada

Abstract

Last year we presented analysis on our project examining the neural networks involved in learning a new ballet to a novel piece of music over 8 months. We scanned dancers up to four times using fMRI along this time period. To date, we have now scanned 11 professional dancers from the National Ballet of Canada and twelve controls. All subjects visualized dancing to a one-minute piece of music during an 8 minute scan. Five epochs of one-minute of music from the ballet were alternated with 30-second fixation periods. Subjects were asked to visualize dancing their part while listening to the music. For more details of the training and performances see our IMRF abstract (DeSouza and Bar, 2012; DOI: 10.1163/187847612X646677). Results revealed a significant increase of BOLD signal, across the sessions in a network of brain regions including bilateral auditory cortex and supplementary motor cortex (SMA) over the first three imaging sessions, but a reduction in the fourth session at 34-weeks. This reduction in activity was not observed in basal ganglia (caudate nucleus). Our results suggest that as we learn a complex motor sequence in time to music, neuronal activity increases until 7 weeks and then decreases by 34-weeks, possibly a result of overlearning. Our findings may also highlight the unique role of basal ganglia regions in the learning of motor sequences. We now aim to use these functional regions of activation as seed regions to explore structural (DTI) and functional connectivity analysis.

Keywords

Plasticity, learning, motor control, auditory, music, dance

Abstract from the 14th International Multisensory Research Forum, The Hebrew University of Jerusalem, Israel, 2013.

* Contact author: rachelbar@gmail.com

 DOI:10.1163/22134808-000S0042

BRILL Multisensory Research 26 Supplement (2013) 65 brill.com/msr

Oral Presentation

Spontaneous gestures modulate speech processing through phase resetting of delta–theta neural oscillations

Emmanuel Biau [1,*], **Salvador Soto-Faraco** [1], **LLuís Fuentemilla** [2] and **Ruth de Diego Balaguer** [2,3]

[1] Multisensory Research Group, Universidad Pompeu Fabra de Barcelona, Spain
[2] Bellvitge Research Biomedical Institute (IDIBELL), Cognition and Brain Plasticity Unit, Spain
[3] Institució Catalana de Recerca i Estudis Avançats (ICREA), Spain

---

Abstract

In everyday life conversations, speakers often accompany verbal utterances with spontaneous gestures (beats) to emphasize certain parts of the utterance. Such affect is thought to be driven by the temporal alignment between visual (gestures) and auditory (speech) inputs. However, the brain mechanisms sustaining this ability remain unknown. We hypothesize that beats, usually preceding word onset by 200 ms, drive the auditory system into an 'optimal state', enhancing its integration of ensuing acoustic information. In this study we tested whether beat gestures reset the phase of ongoing neural oscillations and align neural activity into the high excitability phase at the arrival of the relevant auditory input. Electroencephalographic (EEG) recordings were acquired while participants watched a continuous, naturally recorded political discourse. The phase-locking index (PLI) was calculated from the EEG time locked to the onset of equivalent words appearing naturally in the discourse, sometimes preceded by beat gestures and sometimes not. We observed an increase in phase-locking at the delta–theta frequency range (2–6 Hz) from around 200 ms before word-onset to 200 ms post word-onset, when words were accompanied with a beat gesture compared to audio alone. Furthermore, this increase in phase-locking, most noticeable at fronto-central electrodes, was not accompanied by an increase in power in the same frequency range, confirming the oscillatory-based nature of this effect. These results suggest that beat gestures are used as robust predictive information capable to tune neural oscillations to the optimal phase for auditory integration of relevant parts of the discourse during natural speech processing.

Keywords

Speech, gestures, EEG, oscillations

Abstract from the 14th International Multisensory Research Forum, The Hebrew University of Jerusalem, Israel, 2013.

[*] Contact author: emmanuel.biau@gmail.com

DOI:10.1163/22134808-000S0043

BRILL Multisensory Research 26 Supplement (2013) 66 brill.com/msr

Oral Presentation

Touching bouba, hearing kiki. Image resolution and sound symbolism in visual-to-auditory sensory substitution

Dave J. Brown [1,*] **and Michael Proulx** [2]

[1] Queen Mary University of London, UK
[2] University of Bath, UK

Abstract

The Bouba/Kiki effect involves non-arbitrary mapping between visual shape and speech sounds (Ramachandran and Hubbard, 2001). We presented two groups of participants with both tactile 'bouba and kiki's' and their sonified soundscapes. One group trained in the use of a visual-to-auditory sensory substitution device (SSD), and the other a control group that received no training. Participants in the trained condition learned to make associations between visual and tactile objects and their soundscapes using the SSD. In the test phases the objects were categorized by image resolution and temporal duration. Even though higher resolution images were possible (maximum theoretical resolution of 176 × 144 pixels), successful object discrimination was significant down to 8 × 8 pixels for both tactile-to-auditory and visual-to-auditory conditions. Also, for duration, a highly significant difference favouring the 'short' duration category was found for the tactile-to-auditory condition.

Next, participants were tested with the Bouba/Kiki stimuli. Each group was tested with the tactile shape test. Participants in each condition primarily chose the expected tactile shape (88% trained condition; 83% control), consistent with that chosen visually. Next each group listened to SSD soundscape versions of the stimuli. Here only 41% of the naïve control participants chose the expected soundscape. In contrast, 76.5% of the trained participants selected the expected soundscape that was consistent with that chosen visually (all selected the expected visual–verbal association).

The results show that amongst naive users of SSD's shape discrimination can be made using very basic object features and non-arbitrary cross-modal mappings are apparent after just a basic training regime.

Keywords

Sensory substitution, perceptual learning, sound symbolism

Reference

Ramachandran, V. S. and Hubbard, E. M. (2001). Synaesthesia: A window into perception, thought and language, *Journal of Consciousness Studies* **8**(12), 3–34.

Abstract from the 14th International Multisensory Research Forum, The Hebrew University of Jerusalem, Israel, 2013.

* Contact author: d.j.brown@qmul.ac.uk

 DOI:10.1163/22134808-000S0044

BRILL Multisensory Research 26 Supplement (2013) 67–68 brill.com/msr

Withdrawn Presentation

Does bigger mean louder? Crossmodal congruency and memory judgment

Lionel Brunel *

Epsylon Laboratory, Montpellier 3 University, Montpellier, France

Abstract

Authors have shown that the relation between the perceptual components of the stimulus influences participants' performance during perceptual categorization task (e.g., speeded classification task, Gallace and Spence, 2006; secondary occurrence judgment, Parise and Spence, 2009). For instance, participants are faster at identifying the size of the stimulus when it is accompanied by a congruent tone (e.g., a small circle presented with a low-pitched tone, see Gallace and Spence, 2006; Evans and Treisman, 2010). This mutual relationship has been formalized as cross-modal congruency and could be observed at both semantic and perceptual levels (for a review see Spence, 2011). However, the role of this cross-modal relation during memory judgment is relatively less established. Brunel and collaborators (2010) shown that during typical size judgment from memory, participants automatically simulate auditory characteristics associated with an object. In their paper, they argued that large objects are more likely to simulate sound, as large objects are often associated with loud noise (i.e., cross-modal correspondence between size and loudness). Here we aimed at demonstrating that during typical size judgment from memory, participants could be influenced by the cross-modal congruency between the typical size of a concept and loudness of an accompanying tone. In our experiment, participants had to judge concept-pictures according to their typical size (i.e., small or large objects). Orthogonally from size manipulation, each picture could be accompanied by a sound (either congruent or incongruent) or not. We found that participants were faster at judging typical object-size in congruent condition compared to the other conditions. This result brings support to Brunel and collaborators' (2010) statement and provides a piece of evidence toward a cross-modal correspondence between size and loudness.

Abstract from the 14th International Multisensory Research Forum, The Hebrew University of Jerusalem, Israel, 2013.

* Contact author: lionel.brunel@univ-montp3.fr

 DOI:10.1163/22134808-000S0045

References

Brunel, L., Lesourd, M., Labeye, E. and Versace, R. (2010). The sensory nature of knowledge: Sensory priming effect in semantic categorization, *The Quarterly Journal of Experimental Psychology* **63**, 955–964. DOI:10.1080/17470210903134369

Evans, K. K. and Treisman, A. (2010). Natural cross-modal mappings between visual and auditory features, *Journal of Vision* **10**, 1–12.

Gallace, A. and Spence, C. (2006). Multisensory synesthetic interactions in the speeded classification of visual size, *Perception & Psychophysics* **68**, 1191–1203.

Parise, C. V. and Spence, C. (2009). 'When birds of a feather flock together': Synesthetic correspondences modulate audiovisual integration in non-synesthetes, *PLoS ONE* **4**(5), e5664. DOI:10.1371/journal.pone.0005664

Spence, C. (2011). Crossmodal correspondences: A tutorial review, *Attention, Perception, & Psychophysics* **73**, 971–995. DOI:10.3758/s13414-010-0073-7

BRILL Multisensory Research 26 Supplement (2013) 69–70 brill.com/msr

Oral Presentation

Pre-attentive processing of audition and touch: A high-density electrical mapping study of multisensory mismatch negativity

John S. Butler *, **Sophie Molholm, Ian C. Fiebelkorn, Manuel R. Mercier and John F. Foxe**

The Sheryl and Daniel R. Tishman Cognitive Neurophysiology Laboratory, Department of Pediatrics, Albert Einstein College of Medicine, Bronx, NY, USA

Abstract

Human and animal neurophysiological studies have shown that the auditory and tactile systems interact both early in time, and early in the cortical processing hierarchy. The question remains as to the functional significance of these multisensory interactions that are observed prior to 100 ms post-stimulus onset in both auditory and tactile cortices. We sought to test the hypothesis that this cross-talk occurs during preattentive sensory processing. The mismatch negativity (MMN), an electrophysiological response associated with the pre-attentive detection of a change within the context of a regular pattern of stimulation, served as the dependent measure. High-density electrophysiological recordings were made in healthy adults who viewed a silent movie while they were presented a stream of blocked tactile, auditory or audio-tactile stimuli consisting of frequent and infrequent stimuli. The mismatch waveforms were calculated by subtracting the frequent from the infrequent evoked potential. In Experiment 1 and 2, we investigated the cross sensory interactions to duration and frequency change, respectively. For both experiments we measure tactile, auditory and audio-tactile MMN. The scalp data showed fronto-central negative topographies for both the frequency and duration multisensory MMNs, which is typical of auditory alone mismatches. Cross-sensory interactions between the sensory channels were assessed by testing for non-linearity of the multisensory MMNs when compared to the sum of the unisensory MMNs. In the duration change paradigm the multisensory mismatch showed no cross sensory interaction between audition and touch, suggesting that duration is processed in sensory specific cortices. In contrast, in the frequency change paradigm the multisensory mismatch exhibited non-linear interactions beginning at ~180 ms, and was consistent with preattentive cross-talk among the tactile and auditory temporal frequency processing streams. Our results suggest that different networks might underlie the processing of duration whereas there

---

Abstract from the 14th International Multisensory Research Forum, The Hebrew University of Jerusalem, Israel, 2013.

* Contact author: john.butler@einstein.yu.edu

 DOI:10.1163/22134808-000S0046

is cross-sensory interaction at the sensory perceptual level of audition and tactile for frequency processing.

Keywords
Auditory–somatosensory, EEG

BRILL Multisensory Research 26 Supplement (2013) 71 brill.com/msr

Oral Presentation

The effect of deafness and musical training on perception of space

Zaira Cattaneo [1,*], **Tomaso Vecchi** [2], **Carlo Cecchetto** [1] **and Costanza Papagno** [1]

[1] Department of Psychology, University of Milano-Bicocca, Milano, Italy
[2] Department of Brain and Behavioral Sciences, University of Pavia, Italy

---

Abstract

Individuals tend to show a slight but systematic leftward bias when asked to bisect a line in ether the visual or haptic modality, a tendency known as pseudoneglect (Bowers and Heilman, 1980) and that is likely to reflect a right-hemisphere dominance in spatial attention. Interestingly, when asked to bisect visual lines musicians have been found to be more accurate than non musicians and also to show an opposite slight rightward bias ('minineglect') (Patston *et al.*, 2006). Here we show for the first time that deaf individuals are also more accurate than hearing individuals in bisecting visual lines, and that they show a slight tendency to bisect to the right of the true center, resembling the pattern of musicians. Hence, paradoxically, musical training and sign language seems to affect the way peripersonal space is represented in a similar vein. Interestingly, we observed the same pattern in the haptic modality: when asked to bisect haptic lines, the musicians we recruited were more accurate than non musicians, and showed a slight tendency to bisect to the right of the true midpoint. Deaf individuals showed a more variable pattern depending on the hand used to bisect the line: with the left hand, a clear rightward bias emerged. Overall, our data show that deafness and intense music training may have similar effects on visuo-spatial attention.

Keywords

Space representation, line bisection, deafness, music, visual, haptic

---

Abstract from the 14th International Multisensory Research Forum, The Hebrew University of Jerusalem, Israel, 2013.

* Contact author: zaira.cattaneo@unimib.it

DOI:10.1163/22134808-000S0047

BRILL Multisensory Research 26 Supplement (2013) 72 brill.com/msr

Oral Presentation

Visual experience differentially impacts the egocentric and allocentric coding of touch and motor sequence

Virginie Crollen [1,2,*], **Geneviève Albouy** [2], **Franco Lepore** [2] and **Olivier Collignon** [2,3]

[1] Université Catholique de Louvain, Belgium
[2] Université de Montréal, Montreal, Canada
[3] University of Trento, Italy

Abstract

Touch processing and motor execution both rely on two different spatial frames of reference: a representation built upon egocentric coordinates and a goal-directed representation built upon allocentric coordinates. If early visual deprivation impairs the use of an external frame of reference for touch, the question arises as to whether visual experience also affects the allocentric representation of space in the motor field. To examine this issue, early blind (EB) and sighted control participants (SC) took part in two experiments. In the first, targeting touch processing, participants were required to determine, either with their hands in parallel or crossed over the body midline, the temporal order of two tactile stimuli, one applied to each hand. Crossing the hands led to a significant decrement in performance in SC but did not affect EB, as also demonstrated by Röder and collaborators (2004), suggesting an absence of automatic external remapping of touch in this population. In the second experiment, participants were trained to perform a sequence of five fingers movements. After the training session, participants were tested on their ability to produce, with the same hand, but with the keypad turned upside down, the learned (egocentric condition) or the mirror sequence (allocentric condition). In this task, EB and SC both experienced significant transfer of motor sequence knowledge irrespective of whether the representation of the sequence was allocentric or egocentric. These results therefore demonstrate that early visual experience differentially impacts the egocentric and allocentric coding of touch and motor sequence.

Keywords

Spatial representation, motor sequence, blindness

Reference

Röder, B., Rösler, F. and Spence, C. (2004). Early vision impairs tactile perception in the blind, *Current Biology* **14**, 121–124.

Abstract from the 14th International Multisensory Research Forum, The Hebrew University of Jerusalem, Israel, 2013.

* Contact author: virginie.crollen@uclouvain.be

 DOI:10.1163/22134808-000S0048

BRILL Multisensory Research 26 Supplement (2013) 73 brill.com/msr

Oral Presentation

Audio-tactile crossmodal correspondences

Ophelia Deroy [1,*], Irene Fasiello [2], Vincent Hayward [2] and Malika Auvray [3]

[1] Centre for the Study of the Senses, University of London, UK
[2] ISIR, CNRS UMR 7222, France
[3] LIMSI, CNRS UPR 3251, France

Abstract

We investigated some spontaneous crossmodal correspondences between audition and touch both in blind and sighted people. In four experiments, we tested the interactions between the direction of tactile movement (proximal–distal *vs.* distal–proximal movement on the fingertip) and change in auditory frequency (increasing *vs.* decreasing pitch). We measured the compatibility effect between congruent stimuli (proximal–distal tactile movements and increasing pitch, or distal–proximal tactile movement and decreasing pitch) and incongruent stimuli (i.e., the reverse association). The selective attention method, commonly used to test crossmodal correspondences, requires participants to focus on tactile or auditory signals while ignoring the other one presented simultaneously. The results with this method did not reveal any significant compatibility effect. However, a variant of the implicit association task (IAT, e.g., Parise and Spence, 2012) that relies on associations in the response buttons did reveal a significant compatibility effect. This effect was similar in the conditions where the arm was placed vertically and horizontally, that is whether or not the distal–proximal tactile movement corresponded to the free movement of an object subjected to gravity. Finally, in the IAT protocol, similar effects were obtained in blind and in sighted people, i.e., a crossmodal correspondence effect was obtained independently of the arm's position. These results have methodological implications for the testing of crossmodal correspondences and for the design of sensory substitution devices. They indeed demonstrate the relevance of using spontaneous crossmodal correspondences, and not just arbitrary associations, in order to code aspects of the original signals in conversion systems for blind people.

Keywords

Crossmodal correspondences, audition, touch, implicit association task

Reference

Parise, C. V. and Spence, C. (2012). Audiovisual crossmodal correspondences and sound symbolism: a study using the implicit association test, *Experimental Brain Research* **220**, 319–333.

Abstract from the 14th International Multisensory Research Forum, The Hebrew University of Jerusalem, Israel, 2013.

* Contact author: Ophelia.Deroy@london.ac.uk

 DOI:10.1163/22134808-000S0049

BRILL Multisensory Research 26 Supplement (2013) 74 brill.com/msr

Oral Presentation

Sensitivity of temporal order judgments with repeated stimuli

Massimiliano Di Luca * **and Alice R. Fawdrey**

University of Birmingham, UK

Abstract

Precision in bimodal temporal order judgments is mainly limited by the variability of a central decision mechanism (Sternberg and Knoll, 1973). Ley *et al.* (2009) have shown that precision could be increased if two bimodal stimuli are presented contemporarily. Here we want to investigate whether such an increase in precision can be obtained if the same bimodal stimulus is repeated at a later time.

Participants reported the temporal order of an audiovisual pair, or multiple identical pairs presented in a regular sequence (ISI 400 ms). Trials with different sequence lengths and asynchrony were presented interleaved within one block. Sensitivity increases if two audiovisual pairs are presented as compared to only one. Such an increase is indistinguishable from statistical optimality. Surprisingly however, sensitivity with three and four repeated audiovisual stimuli is suboptimal and in fact identical to the sensitivity obtained with two stimuli.

In a second experiment, two audiovisual pairs were presented in succession with a slight difference in asynchrony (60 ms). Sensitivity is still close to optimal in this case and the point of subjected simultaneity indicates that the two audiovisual stimuli have been given similar weight in the judgment.

These results indicate that precision in temporal order judgments can benefit from stimulus repetition, but the improvement is limited to two repetitions. Such a limit can be related to the temporal integration characteristics of the decision mechanism.

Keywords

Simultaneity perception, psychophysics, Bayesian modelling, temporal order judgments

References

Ley, I., Haggard, P. and Yarrow, K. (2009). Optimal integration of auditory and vibrotactile information for judgments of temporal order, *J. Exp. Psychol. Human* **35**, 1005–1019.

Sternberg, S. and Knoll, R. (1973). The perception of temporal order: Fundamental issues and a general model, in: *Attention and Performance IV*, S. Kornblum (Ed.), pp. 629–685. Academic Press, New York.

Abstract from the 14th International Multisensory Research Forum, The Hebrew University of Jerusalem, Israel, 2013.

* Contact author: m.diluca@bham.ac.uk

 DOI:10.1163/22134808-000S0050

BRILL Multisensory Research 26 Supplement (2013) 75–76 brill.com/msr

MULTISENSORY RESEARCH

Oral Presentation

Behavioral and neurophysiological correlates of sight restoration after longstanding visual deprivation

Giulia Dormal [1,2,*], Franco Lepore [1], Mona Harissi-Dagher [3], Geneviève Albouy [4], Alan Tucholska [5], Armando Bertone [6], Bruno Rossion [2] and Olivier Collignon [7,8]

[1] Centre de Recherche en Neuropsychologie et Cognition, University of Montreal, Canada
[2] Institut de Psychologie et Institut Neurosciences, University of Louvain, Belgium
[3] Ophtalmology Department, Hôpital Notre-Dame, University of Montreal, Canada
[4] Centre de Recherche de l'Institut Universitaire de Gériatrie, University of Montreal, Canada
[5] Radiology Department, Hôpital Notre-Dame, University of Montreal, Canada
[6] Perceptual Neuroscience Laboratory for Autism and Development, McGill University, Canada
[7] Centre de Recherche du CHU Sainte-Justine, University of Montreal, Canada
[8] Centre for Mind/Brain Science (CIMeC), University of Trento, Italy

Abstract
Visual deprivation leads to the involvement of occipital (visual) regions in the processing of non-visual inputs. These mechanisms of crossmodal plasticity, classically considered compensatory, inevitably raise crucial challenges for sight-restoration. In the current study, we had the unique opportunity to track the behavioral and neurophysiological changes taking place in the occipital cortex of an early visually deprived patient before and after partial sight restoration with Boston Keratoprosthesis. The patient was tested 3 weeks before, as well as 1.5 and 7 months after the surgery with identical behavioral and fMRI tasks. Behavioral tasks consisted of computerized tests evaluating visual acuity, contrast sensitivity, global motion perception, and face categorization and individuation. Each fMRI session comprised 2 visual runs testing the integrity of face-selective and motion-selective networks, and 2 auditory runs investigating the presence of general and specific crossmodal reorganizations. The success of the intervention was witnessed by massive improvements in visual functions. fMRI analyses revealed (1) massive auditory responses in occipital regions, which were still existing 7 months post-surgery, but significantly reduced when compared to pre-surgery; (2) a regain of specific responses to visual motion 7 months post-surgery relative to pre-surgery in early visual regions;

Abstract from the 14th International Multisensory Research Forum, The Hebrew University of Jerusalem, Israel, 2013.

* Contact author: giulia.dormal@uclouvain.be

 DOI:10.1163/22134808-000S0051

(3) no between sessions differences in higher level regions involved in face- (FFA, OFA) and motion- (hMT+/V5) processing suggesting that even rudimentary vision may set the functional tuning of these regions. Finally, we observe massive structural changes in primary occipital cortex as soon as 1.5 months after surgery. Present results are the first to unravel the important neuro-functional and structural changes ensuing sight restoration and are clinically important given the therapeutic advances in sensory-restoration.

Keywords
fMRI, blindness, sight restoration, Boston keratoprosthesis, crossmodal plasticity, face processing, motion processing

BRILL

Multisensory Research 26 Supplement (2013) 77

MULTISENSORY RESEARCH

brill.com/msr

Oral Presentation

Multisensory integration for expert musicians investigated using eye tracking

Veronique Drai-Zerbib * **and Thierry Baccino**

LUTIN CHART Cite Des Sciences et de L'industrie, Paris, France

Abstract

Musical sight-reading requires to process simultaneously multimodal information: visual, auditory, motor. Does the expertise in music rely on an efficient cross-modal integration? This talk investigates this issue with 2 experiments. In the first one, 30 expert and 31 non-expert musicians were required to report whether two successively presented fragments of classical music were same or different. In half the conditions the 61 participants received the fragments in the same modal presentation (visual/visual), in the other half they received the fragments in cross-modal presentation (auditory/visual). Analysis of Response Time and Errors showed that more experienced musicians seemed to be better able to transfer the information from one modality to another. In a second experiment using eye-tracking, 64 participants, 26 expert and 38 non-expert musicians were required also to report whether two successively presented fragments of classical music were same or different but in cross-modal way only. Visual and auditory cues were used to investigate whether a kind of expert memory using retrieval cues was acting for more expert musicians. An accent mark, emphasis placed on a note contributing to the prosody of the musical phrase, was put in a congruent or incongruent way, during the auditory and reading phases. As expected, the analysis of fixations and mistakes validated the hypothesis of modal independence for expert musicians, observed in the first experiment. Moreover analyses validated the cross-modal ability of expert memory, using accent marks as retrieval cues. Results are discussed in terms of amodal memory for expert musicians that can be in support of theoretical work by Ericsson and Kintsch (1995): more experienced performers better integrate knowledge across modalities using retrieval cues.

Keywords

Musical reading, expertise, memory

Reference

Ericsson, K. and Kintsch, W. (1995). Long-term working memory, *Psychological Review* **102**(2), 211–245.

Abstract from the 14th International Multisensory Research Forum, The Hebrew University of Jerusalem, Israel, 2013.

* Contact author: zerbib@unice.fr

DOI:10.1163/22134808-000S0052

BRILL Multisensory Research 26 Supplement (2013) 78 brill.com/msr

Oral Presentation

The effects of a multisensory training for hemianopia on spatial attention

Neil Dundon *, **Martin Maier, Caterina Bertini and Elisabetta Ladavas**

University of Bologna, Italy

Abstract

The aim of the study is to assess the neural mechanisms underlying the recovery of visual function after multisensory stimulation (MS) in hemianopic patients. As a result of a hyperactivation of the intact hemisphere, with concurrent hypoactivation of the damaged hemisphere, hemianopic patients focus their attention on the ipsilesional field and the contralesional field lacks sufficient attentional resources. The unbalanced excitability of the two hemispheres can be restored by a multisensory stimulation, mediated by the collicular-extrastriate pathway, which largely contributes to spatial orienting behaviors. We administered a course of MS treatment (Bolognini *et al.*, 2005) to a sample of seven hemianopic participants for 4 hours daily, over a period of nearly 2 weeks and we collected both behavioural and electrophysiological data, before and after the intervention, in order to index the stimulation effects. The results showed an improvement of visual oculomotor exploration that allowed patients to efficiently compensate for the loss of vision. In addition, amplitudes of P3 event related potentials elicited by a simple visual detection paradigm were significantly reduced after the treatment when the stimuli were presented to the intact field. The behavioural improvement in the hemianopic field, and ERP amplitude reduction in the intact field, can be explained as driven by a shift in spatial attention away from the hyperactivated intact visual field, induced by the intensive multisensory training.

Keywords

Visual field defect, spatial attention, ERP

Reference

Bolognini, N., Rasi, F., Coccia, M. and Làdavas, E. (2005). Visual search improvement in hemianopic patients after audio-visual stimulation, *Brain* **128**(12), 2830–2842.

Abstract from the 14th International Multisensory Research Forum, The Hebrew University of Jerusalem, Israel, 2013.

* Contact author: neildundon@gmail.com

 DOI:10.1163/22134808-000S0053

BRILL Multisensory Research 26 Supplement (2013) 79 brill.com/msr

MULTISENSORY RESEARCH

Oral Presentation

Throwing out the rules: Anticipatory alpha-band oscillatory attention mechanisms during intersensory task-set reconfigurations

John J. Foxe *, **Jeremy W. Murphy and Peirfilippo De Sanctis**
Albert Einstein College of Medicine, Bronx, NY, USA

Abstract
When individuals are required to switch rapidly from execution of one task to another, goal-related task networks and attentional mechanisms are engaged to reconfigure task-specific networks, suppressing activity within circuits responsible for performance of the old task and amplifying preparatory neural processes for the anticipated novel task. That is, competition between two potential task-set configurations must be resolved so that an effective strategy shift can be enacted. In this study, we investigated the neural processes involved in switching between visual and auditory tasks, where both potentially task-relevant sensory inputs were presented simultaneously. As such, the two sensory systems were competing for processing resources. We were specifically interested in the role of oscillatory neural mechanisms in these task-set reconfigurations. We principally examined alpha-band oscillatory activity (10–14 Hz) in the periods preceding switches of task, compared to periods preceding repeats of a given sensory task. The need to switch tasks across the senses resulted in significant behavioral costs (slower and less accurate performing) than when just one task was engaged in for protracted periods of time. Anticipatory alpha-band suppressive mechanisms, in the form of substantially amplified oscillatory synchronizations, were implicated in the successful switching of attention to the auditory stream. In contrast, vigorous alpha-band desynchronization characterized periods preceding switches to the visual domain. The major cortical generators of this alpha-band activity were found in parieto-occipital and fronto-central regions, suggesting involvement of the well-known fronto-parietal attention network in resolving competition between two competing sensory tasks.

Keywords
Attention, competition, oscillations, alpha-band, task-switching

Abstract from the 14th International Multisensory Research Forum, The Hebrew University of Jerusalem, Israel, 2013.

* Contact author: john.foxe@einstein.yu.edu

 DOI:10.1163/22134808-000S0054

BRILL Multisensory Research 26 Supplement (2013) 80–81 brill.com/msr

Oral Presentation

Learning regularities from different sensory modalities: Evidence for an a-modal learning mechanism

Arit Glicksohn *

Department of Psychology, The Hebrew University of Jerusalem, Israel

Abstract

How do we implicitly learn regularities from different modalities? Implicit learning refers to people's ability to learn regularities in the environment automatically, unintentionally and without awareness. One central question is whether learning occurs separately within modalities (e.g., vision, audition) or whether learning is a-modal. Previous studies largely support separable modality-specific learning mechanisms. We explored this question using two prominent learning paradigms: statistical learning (SL), referring to learning of statistical regularities between elements such as transitional probabilities, and artificial grammar learning (AGL), referring to learning the underlying grammar of a set of exemplars. In Experiment 1 participants were familiarized with a structured stream of elements composed of reoccurring triplets, and subsequently were tested on their familiarity of these triplets. Learning occurred for visual, auditory and audiovisual triplets. Moreover, learning of the audiovisual triplets was significantly larger compared to learning of unimodal triplets. Similarly, using an AGL task, in Experiment 2 participants were familiarized with a set of sequences all adhering to a grammar. Participants could subsequently classify novel sequences as grammatical or not for both multimodal and unimodal sequences, with learning superior for multimodal compared to unimodal learning. In Experiment 3 we used a modified AGL task offering an alternative explanation for previous studies indicating modality-specific learning mechanisms. Two main findings are noted: (a) Implicit learning is amodal, learning both unimodal and multimodal regularities, and (b) multimodal is higher compared to unimodal learning. Together, these findings support the existence of an a-modal learning mechanism(s) sensitive to multimodal information.

Keywords

Implicit learning, artificial grammar learning, statistical learning, multimodal learning

Abstract from the 14th International Multisensory Research Forum, The Hebrew University of Jerusalem, Israel, 2013.

* Contact author: arit.glicksohn@mail.huji.ac.il

DOI:10.1163/22134808-000S0055

References

Aslin, R. N. and Newport, E. L. (2012). Statistical learning: From acquiring specific items to forming general rules, *Current Directions in Psychological Science* **21**, 170–176. DOI:10.1177/0963721412436806

Cohen, A., Ivry, R. I. and Keele, S. W. (1990). Attention and structure in sequence learning, *Journal of Experimental Psychology: Learning, Memory, and Cognition* **16**, 17–30. DOI:10.1037//0278-7393.16.1.17

Conway, C. M. and Christiansen, M. H. (2006). Statistical learning within and between modalities, *Psychological Science* **17**, 905–912. DOI:10.1111/j.14679280.2006.01801.x

Fiser, J. and Aslin, R. N. (2002). Statistical learning of higher-order temporal structure from visual shape sequences, *Journal of Experimental Psychology: Learning, Memory, and Cognition* **28**, 458–467. DOI:10.1037//0278-7393.28.3.458

Kirkham, N. Z., Slemmer, J. A. and Johnson, S. P. (2002). Visual statistical learning in infancy: evidence for a domain general learning mechanism, *Cognition* **83**, B35–B42. DOI:10.1016/S0010-0277(02)00004-5

Reber, A. (1967). Implicit learning of artificial grammars, *Journal of Verbal Learning and Verbal Behavior* **6**, 855–863.

BRILL Multisensory Research 26 Supplement (2013) 82 brill.com/msr

Oral Presentation

Stop, blink and listen

Keren Haroush and Shaul Hochstein *

Hebrew University, Jerusalem, Israel and Harvard Medical School, Cambridge, MA, USA

Abstract

What happens to cross-modal perceptual resources when task performance is predictably poor? This question is central for understanding the efficiency of attention allocation among concurrent tasks in different modalities and for assigning multi-tasking operations in real-life situations (such as texting while driving a car). General load theories propose that attention allocation may be a zero-sum game so that the more difficult one task, the poorer performance on a second. Consistent with this view, having detected a first target, observers often fail to report a closely following second, the Attentional-Blink. What happens when different senses are involved? We tested the novel prediction that attention can be shifted from tasks that will probably fail anyway, to improve performance of a concurrent, more promising task. To test this, participants performed auditory pitch discrimination while engaged in detecting two successive visual targets in an Attentional-Blink paradigm. We found that, rather than suffering from general system load, auditory performance was significantly superior during the Attentional-Blink. We conclude that when visual task failure was predicted, resources were shifted to the auditory system. These findings suggest presence of a highly sophisticated attention–allocation mechanism that computes likelihood-of-success *versus* cost for each task on a moment-to-moment basis.

Abstract from the 14th International Multisensory Research Forum, The Hebrew University of Jerusalem, Israel, 2013.

* Contact author: shaulhochstein@gmail.com

 DOI:10.1163/22134808-000S0056

BRILL
Multisensory Research 26 Supplement (2013) 83–84
MULTISENSORY RESEARCH
brill.com/msr

Oral Presentation

At the tip of the fingers: The role of orientation and position on the reference frames used to interpret ambiguous stimuli presented to the hand

Jess Hartcher-O'Brien [1,*], **Charles J. Spence** [2] **and Malika Auvray** [1]
[1] LIMSI CNRS, France
[2] University of Oxford, UK

Abstract
Coding an object or events position in the environment involves recruitment of multiple reference frames or coordinate systems. Generally retino-, head-, trunk-, arm and object-centred representations have been shown to influence how we estimate relative object position. Visual information can primarily be coded in retinotopic or spatiotopic frames while tactile information does not provide such a one-to-one mapping between position in space and receptor activation. The question asked here is how tactile information presented to the hand, which is able to change its orientation in space and body coordinates, influences interpretation of tactile letter stimuli. As letters are generally perceived visually, when interpreting a letter traced on the skin, one might assume that the perceptual system defaultly employs a visual frame of reference, or at least a simple egocentric perspective. However, previous findings suggest that for cutaneous tracing the perspective taken is influenced in a complex manner by factors such as location, as opposed to receptor type, limb orientation, location of stimulation relative to other body parts and object coordinates (Parsons and Shimojo, 1987). In three experiments we explored the initial interpretation of letter stimuli d, b, p, & q when presented to the hand and fingertips. The stimuli are ambiguous along left/right and top/bottom axes and are therefore ideal for inferring reference frame adopted. The current findings suggest that orientation and position of the hand relative to the body, alters the reference frame selected when interpreting tactile letter stimuli. The results are consistent with a multiple reference frame model of information processing and allow us to disentangle the elements involved in perception of cutaneous stimulation at our fingertips.

Keywords
Tactile perception, reference frames, ambiguous stimuli, perspective, fingertip

Abstract from the 14th International Multisensory Research Forum, The Hebrew University of Jerusalem, Israel, 2013.
* Contact author: hartcher@limsi.fr

DOI:10.1163/22134808-000S0057

Reference

Parsons, L. M. and Shimojo, S. (1987). Perceived spatial organization of cutaneous patterns on surfaces of the human body in various positions, *Journal of Experimental Psychology: Human Perception & Performance* **13**, 488–504.

BRILL Multisensory Research 26 Supplement (2013) 85 brill.com/msr

Oral Presentation

Multisensory flexibility within a perceptual system reorganized by crossmodal plasticity

Benedetta Heimler *, **Francesca Baruffaldi, Roberta Taddei and Francesco Pavani**

University of Trento, Italy

Abstract

The widespread intuition that sensory deprivation should enhance processing in the intact modalities has been primarily tested in unisensory contexts. This approach has the limit of not probing for potential changes in the interaction between intact modalities, ultimately not allowing to conclude whether documented unisensory advantages always dominate behavior. Here we examined whether deafness modifies the interactions between vision and touch, and to what extent enhanced processing of peripheral visual events — repeatedly documented in unisensory studies of deafness — dominates behavior also in a multisensory context in which vision is entirely task-irrelevant. Nine hearing and seven early-deaf adults performed two visuo-tactile tasks. In one task, participants responded to the elevation of a tactile stimulus while ignoring a concurrent visual stimulus (central or peripheral). The distractor was spatially congruent or incongruent with the target. The other task was reversed (i.e., respond to vision, ignore touch). Visuo-tactile interference emerged for all participants, revealing similar multisensory processing in the two groups. However, when vision was task relevant, deaf people were distracted by touch less than hearing controls when visual targets were peripheral. This reveals the expected enhanced processing for peripheral visual stimuli in deafness. Strikingly, when vision was task irrelevant (i.e., respond to touch) comparable visuo-tactile interference emerged in the two groups for both visual distractor eccentricities. These findings show that enhanced processing of peripheral visual information in deafness can be modulated as a function of task-demands, revealing a remarkable multisensory flexibility within a perceptual system reorganized by crossmodal plasticity.

Keywords

Deafness, cross-modal plasticity, multisensory interference

Abstract from the 14th International Multisensory Research Forum, The Hebrew University of Jerusalem, Israel, 2013.

* Contact author: benedetta.heimler@unitn.it

DOI:10.1163/22134808-000S0058

BRILL Multisensory Research 26 Supplement (2013) 86 brill.com/msr

Oral Presentation

Selective attention determines the direction of audiovisual temporal recalibration

Nara Ikumi [1,*] **and Salvador Soto-Faraco** [1,2]

[1] Center for Brain and Cognition (CBC), Universitat Pompeu Fabra, Spain
[2] Institució Catalana de Recerca i Estudis Avançats (ICREA), Spain

---

Abstract

Coherent perception of multimodal events requires the adjustment of temporal discrepancies between information from different sensory modalities. Temporal recalibration by adaptation to asynchrony has been proposed as a mechanism to compensate for these timing differences. However, most of the studies to date have examined recalibration on isolated crossmodal events. Considering the rich complexity of everyday life sensory environments, it seems crucial to investigate how the brain selects which stimuli, of all possible events falling within the same temporal window, should be adjusted in time. Here, we hypothesize that attention might provide an effective filter to help resolve which stimuli are selected when multiple events compete for recalibration. To test this we used an audiovisual recalibration paradigm in which two alternative asynchronies were present during the adaptation. The exposure phase consisted of a repetitive sequence of flash–tone–flash, and we manipulated the direction of attention to one of two visual events. In the test phase, following exposure, we found a shift in the point of subjective simultaneity (PSS) depending on the direction of attention, in otherwise physically identical adaptation conditions. A follow up experiment where temporal attention was deployed at alternate temporal positions revealed that recalibration aftereffects are also driven by stimulus properties. These results suggest that attention to specific stimuli promotes certain audiovisual pairings events to be combined and subsequently adjusted in time. Furthermore, we speculate that the resolution of recalibration in complex scenes involves the orchestration between top-down selection mechanisms and stimulus-driven processes.

Keywords

Temporal recalibration, audiovisual simultaneity, attention

Abstract from the 14th International Multisensory Research Forum, The Hebrew University of Jerusalem, Israel, 2013.

* Contact author: nara.ikumi@gmail.com

DOI:10.1163/22134808-000S0059

BRILL Multisensory Research 26 Supplement (2013) 87–88 brill.com/msr

Oral Presentation

Cerebral encoding of structures from different sensory sources

Corinne Jola *

NeuroSpin Centre, INSERM/CEA, France

Abstract

Recent neuroimaging studies have investigated cortical activity in response to passive exposure to long segments of visual and/or auditory stimuli (e.g., Hasson *et al.*, 2004; Jola *et al.*, 2013; Speer *et al.*, 2007). The transition from short to long stimulus material enables studying cognitive processes that are closer to real life. Moreover, processes involved in perception of structure in time do ultimately require stimuli of extended duration. Despite the challenges that such complex but ecologically more valid stimuli pose, the results are promising. I will thus present recent findings of how long stimuli are processed in the case of dance. Dance is a multisensory stimulus; as it is generally performed to music. We found evidence of enhanced synchronisation of neuronal activity in participants' frontal brain areas when the dance recordings were edited (and thus more directed) compared to when they were presented as a single shot recording. This is in support of the editing hypothesis, namely, that visually presented stimuli require directing spectators' attention in order to elicit enhanced shared responses (see Jola *et al.*, 2013). Yet a number of additional studies showed that spectators perceive dance structures in very individual ways (e.g., deLahunta and Barnard, 2005; Pollick *et al.*, 2012). However, this heterogeneity is reduced if the spectators have a shared practice: we found evidence for enhanced synchronisation in sensorimotor and frontal areas in response to watching a single shot dance video, under the condition that the participants have physically learned the movement before they passively observed them in the scanner (Jola *et al.*, in prep.). This finding suggests that for shared motor resonance, a predisposition of a consolidated structure is important. On the premise of a multisensory structure encoding mechanism, we thus currently investigate the cerebral encoding of dance structures in analogy to language and music processing (Pallier *et al.*, 2011). I will outline the experimental challenges an ecologically valid study of the moving (i.e., dancing) human body poses and present our preliminary results.

Keywords

Action observation, fMRI, audiovisual composition

Abstract from the 14th International Multisensory Research Forum, The Hebrew University of Jerusalem, Israel, 2013.

* Contact author: Corinne.Jola@cea.fr

 DOI:10.1163/22134808-000S0060

References

Hasson, U., Nir, Y., Levy, I., Fuhrmann, G. and Malach, R. (2004). Intersubject synchronization of cortical activity during natural vision, *Science* **303**, 1634–1640.

Herbec, A., Kauppi, J. P., Jola, C., Haodan, T., Tohka, J. and Pollick, F. E. (in prep.). Inter-subject synchronization while viewing edited *versus* non-edited videos of a dance performance.

Jola, C., McAleer, P., Grosbras, M. H., Love, S. A., Morison, G. and Pollick, F. E. (2013-in press). Uni- and multisensory brain areas are synchronised across spectators when watching unedited dance, *i-Perception*. dx.doi.org/10.1068/i0536IMRF

Jola, C., Gazzola, V. and Keysers, Ch. (in prep.). Synchronised brain activity during action observation across expert spectators.

Katie, N., Glowinski, D., Murphy, H., Jola, C., McAleer, P., Darshane, N., Penfield, K., Camurri, A. and Pollick, F. E. (in prep.). Brain mechanisms of action understanding revealed by watching extended dance solos.

De Lahunta, S. and Barnard, P. (2005). What's in a phrase?, in: *Tanz im Kopf*, J. Birringer and J. Fenger (Eds). LIT Verlag, Hamburg.

Pallier, C., Devauchelle, A. D. and Dehaene, S. (2011). Cortical representation of the constituent structure of sentences, *Proc. Natl. Acad. Sci.* **108**, 2522–2527.

Pollick, F., Noble, K., Darshane, N., Murphy, H., Glowinski, D., McAleer, P., Jola, C., Penfield, K. and Camurri, A. (2012). Using a novel motion index to study the neural basis of event segmentation. Conference Proceedings, *i-Perception*, **4**, 225.

Reason, M., Reynolds, D., Kay, R., Jola, C., Grobras, M.-H. and Pollick, F. E. (in prep.). Spectators' aesthetic experience of sound and movement in dance performance: a multi-methodological investigation.

Speer, N. K., Zacks, J. M. and Reynolds, J. R. (2007). Human brain activity time-locked to narrative event boundaries, *Psychol. Sci.* **18**, 449–455.

BRILL Multisensory Research 26 Supplement (2013) 89 brill.com/msr

Oral Presentation

Phase encoding of perceived events timing

Anne Kösem * **and Virginie van Wassenhove**

INSERM-CEA Cognitive Neuroimaging Unit, France

Abstract

Temporal coincidence is a crucial feature for audiovisual (AV) integration, and recent neurophysiological findings provide evidence for the role of oscillations in the temporal binding of information. Specifically, low-frequency oscillatory phase alignment between visual and auditory cortices has been proposed to support AV integration in time (Lakatos *et al.*, 2008; Schroeder and Lakatos, 2009). Here we hypothesized that if phase coincidence was critical for perceiving synchrony between auditory and visual events, shifts in the phase should be commensurate with the subjective perception of AV events. To test this, participants underwent AV temporal recalibration (Fujisaki *et al.*, 2004) while being recorded with magnetoencephalography (MEG). Temporal recalibration consisted in adapting subjects to desynchronized AV events in order to induce perceptual shifts of AV temporal order. Each recalibration period was followed by an assessment of participant's temporal order threshold. Stimuli during recalibration were presented at an average rate of 1 Hz leading to a prominent tagging in both auditory and visual cortices. Analyses were carried out in source space (dSPM, cortically constrained source orientations) and restricted to sensory cortices. We show that the variability in perceived temporal order is captured by non-stationary 1 Hz entrainment: the preferential phase at this frequency was significantly shifted between the beginning and the end of the recalibration period in sensory cortices. Individuals' perceived simultaneity could be accounted for by systematic shifts in the phase of auditory but not visual neural responses. All together, our results provide a novel physiological index for subjective simultaneity suggesting that auditory cortex recalibrates its timing to the visual spatial anchor.

Keywords

Temporal order, perception, oscillatory entrainment, multisensory, MEG, simultaneity, delta band

References

Fujisaki, W., Shimojo, S., Kashino, M. and Nishida, S. (2004). Recalibration of audiovisual simultaneity, *Nature Neuroscience* **7**, 773–778.

Lakatos, P., Karmos, G., Mehta, A. D., Ulbert, I. and Schroeder, C. E. (2008). Entrainment of neuronal oscillations as a mechanism of attentional selection, *Science* **320**, 110–113.

Schroeder, C. E. and Lakatos, P. (2009). Low-frequency neuronal oscillations as instruments of sensory selection, *Trends in Neurosciences* **32**, 9–18.

Abstract from the 14th International Multisensory Research Forum, The Hebrew University of Jerusalem, Israel, 2013.

* Contact author: kosem.anne@gmail.com

 DOI:10.1163/22134808-000S0061

BRILL

Multisensory Research 26 Supplement (2013) 90

MULTISENSORY RESEARCH

brill.com/msr

Oral Presentation

Nonlinear bimodal integration of on/off visual and auditory signals in awake rat V1

Kuo-Sheng Lee [1,*] **and Chun-I. Yeh** [2]

[1] Department of Psychology, National Taiwan University, Taipei, Taiwan

[2] Department of Psychology, Graduate Institute of Brain and Mind Sciences, Neurobiology and Cognitive Science Center, National Taiwan University, Taipei, Taiwan

Abstract

Detecting changes is a critical feature of sensory processing. In the visual system, light increments and decrements are processed through parallel ON and OFF channels that converge in the primary visual cortex (V1). Similarly in the auditory system, UP/DOWN signals (stimulus onset/offset) are carried by non-overlapping sets of neurons that converge in the primary auditory cortex (A1) (Scholl *et al.*, 2010). A recent study has shown that responses of V1 neurons are greatly suppressed by auditory signals in rodents (Iurilli *et al.*, 2012). However, it is unclear whether auditory signals may modulate ON and OFF channels differently. Moreover, the modulation of V1 neurons may depend on the degree of congruence between auditory and visual stimuli (e.g. UP–ON *versus* DOWN–ON). We addressed these questions by recording from V1 neurons in awake Long–Evans rats. Brief auditory (pure tone) and visual (full field) stimuli were presented simultaneously (0.2 s), and a total of 25 bimodal stimulation conditions (5×5) were used (auditory: mean amplitude: 80 dB; UP/DOWN: +13% and +25%; visual: mean amplitude: 70 cd/m^2; ON/OFF: +35% and +70%). Preliminary results showed that the effect of bimodal stimulation was mostly sub-additive — the bimodal responses were weaker than the linear mathematic combination of the two excitatory unimodal responses. Among the four bimodal conditions, neural responses in the OFF–DOWN condition tend to be the most sub-additive. Overall, these results suggest that a linear combination of A1 and V1 responses cannot explain the complex interaction between ON/OFF channels and UP/DOWN channels.

Keywords

Primary visual cortex, audiovisual integration, rat

References

Iurilli, G., Ghezzi, D., Olcese, U., Lassi, G., Nazzaro, C., Tonini, R., Tucci, V., Benfenati, F. and Medini, P. (2012). Sound-driven synaptic inhibition in primary visual cortex, *Neuron* **73**, 814–828.

Scholl, B., Gao, X. and Wehr, M. (2010). Nonoverlapping sets of synapses drive on responses and off responses in auditory cortex, *Neuron* **65**, 412–421.

Abstract from the 14th International Multisensory Research Forum, The Hebrew University of Jerusalem, Israel, 2013.

* Contact author: b98207071@ntu.edu.tw

DOI:10.1163/22134808-000S0062

BRILL Multisensory Research 26 Supplement (2013) 91 brill.com/msr

Oral Presentation

Sound speeds vision through preparation, not integration

Sander Los [1,*] **and Erik van der Burg** [2]

[1] VU University, Amsterdam, The Netherlands

[2] The University of Sydney, Australia

Abstract

People respond faster to a visual target stimulus when it is accompanied by a task-irrelevant tone. This intersensory facilitation effect is often attributed to multisensory integration, but here we show it to be a reflection of temporal preparation. According to this view, the more rapidly processed tone serves as a warning signal (S1), which initiates preparation for the more sluggish visual target (S2). To test this view, we varied the delay between S1 and S2 in conjunction with the modality of S1 (auditory or visual). For brief delays, responses to S2 were faster when S1 was auditory than when it was visual. Crucially, however, this intersensory facilitation effect disappeared after a simple correction for the difference in S1-detection time, equating the effective preparation period. The obtained audio-visual equivalence shows that sound speeds response to a visual target through preparation, not integration.

Keywords

Intersensory facilitation, temporal preparation, multisensory integration

Abstract from the 14th International Multisensory Research Forum, The Hebrew University of Jerusalem, Israel, 2013.

* Contact author: s.a.los@vu.nl

DOI:10.1163/22134808-000S0063

BRILL
Multisensory Research 26 Supplement (2013) 92
MULTISENSORY RESEARCH
brill.com/msr

Oral Presentation

Early interaction between vision and touch during binocular rivalry

Claudia Lunghi [1,2,*] **and Maria C. Morrone** [3,4]

[1] Department of Neuroscience, Università Degli Studi di Firenze, Italy
[2] Institute of Neuroscience, CNR, Pisa, Italy
[3] Department of Translational Research on New Technologies in Medicine and Surgery, Università di Pisa, Italy
[4] Scientific Institute Stella Maris, Calambrone, Pisa, Italy

Abstract

Growing evidence shows that early sensory areas are involved in cross-modal processing, challenging the traditional view that confined multisensory analysis to high-level temporal and parietal areas. Haptic signals have been shown to influence the dynamics of binocular rivalry, a form of perceptual bistability that engages competition between monocular signals (Lunghi *et al.*, 2010, *Current Biology*, **20**(4), R143–R144). We investigated the role of spatial proximity, type of tactile stimulation (active or passive) and cross-modal attention in promoting fusion between vision and touch during binocular rivalry of orthogonally oriented gratings (±45°, size 2.5°, SF 2 c/cm). By varying the haptic orientation (±7.5°, ±15°, ±30°), we further investigated the orientation selectivity of the interaction. We found that both active exploration and passive tactile stimulation boosted the visual stimulus congruent with the tactile one (engraved grating), both prolonging conscious perception and restoring it from binocular rivalry suppression. The interaction was strictly tuned (less than one octave) to visuo-haptic spatial frequency and orientation. We also found that voluntary attention played a minor role in mediating the interaction, while temporal alignment and spatial proximity between visual and tactile stimuli were necessary for fusion of the cross-sensory signals, this latter result suggesting that the visual and the somatosensory spatial maps are aligned. Taken together, our results points to a very early multisensory interaction site, possibly V1.

Keywords

Cross-modal processing, binocular rivalry, psychophysics

Abstract from the 14th International Multisensory Research Forum, The Hebrew University of Jerusalem, Israel, 2013.
* Contact author: claudia.lunghi@in.cnr.it

DOI:10.1163/22134808-000S0064

BRILL

Multisensory Research 26 Supplement (2013) 93

MULTISENSORY RESEARCH

brill.com/msr

Oral Presentation

(Non)sensory reorganisation following arm amputation

Tamar Makin [1,*], Nicola Filippini [2], Jan Scholz [3,4], Eugene Duff [1], David Henderson Slatere [5], Irene Tracey [6] and Heidi Johansen-Berg [2]

[1] FMRIB Centre, University of Oxford, UK

[2] University of Oxford, UK

[3] FMRIB Centre, Nuffield Department of Clinical Neurosciences, University of Oxford, Oxford OX3 9DU, UK

[4] Mouse Imaging Centre, Hospital for Sick Children, Toronto, Ontario, Canada M5T 3H7

[5] Oxford Centre for Enablement, Nuffield Orthopaedic Centre, Oxford OX3 7HE, UK

[6] The Oxford Centre for Neuroethics, University of Oxford, UK

Abstract

Arm amputation has been shown to result in dramatic unisensory reorganisation within the sensorimotor homunculus. However, little is known about the consequences of sensorimotor deprivation on global brain organisation. We studied the interplay between reorganisation within the primary sensorimotor cortex and reorganisation across large-scale brain networks in arm amputees and controls. Unisensory reorganisation was measured as medial shifts in lip representation in the deprived homunculus using task-based fMRI, reflecting the well-documented invasion of the lip representation into the missing hand area. Network-level reorganisation was assessed as changes in functional connectivity strength between the deprived hand area and fMRI resting-state networks, using independent component analysis. Functional connectivity between the deprived cortex and its network of origin (the sensorimotor network) was reduced in amputees, compared with controls. This was associated with greater coupling of the deprived cortex with the non-sensory default mode network, from which the primary sensorimotor cortex is normally decoupled. The degree of this new (non)sensory coupling was correlated with the degree of unisensory remapping within the sensorimotor homunculus and with decoupling of the deprived hand area from the resting-state sensorimotor network. Our results demonstrate that, following amputation, plasticity is present in, but not restricted to, the sensorimotor network, highlighting the role of sensory deprivation as a driving force for both sensory and non-sensory plasticity.

Keywords

Sensory deprivation, resting states networks, fMRI

Abstract from the 14th International Multisensory Research Forum, The Hebrew University of Jerusalem, Israel, 2013.

* Contact author: tmakin@fmrib.ox.ac.uk

DOI:10.1163/22134808-000S0065

BRILL Multisensory Research 26 Supplement (2013) 94 brill.com/msr

Oral Presentation

Selective attention to sound modulates neural activity in areas of audiovisual integration

Luis Morís Fernández [1,*], **Maya Visser** [2], **César Ávila** [2] **and Salvador Soto-Faraco** [1,3]

[1] Center for Brain and Cognition (CBC), Universitat Pompeu Fabra, Spain
[2] Universitat Jaume I, Spain
[3] Institució Catalana de Recerca i Estudis Avançats (ICREA), Spain

Abstract

Although the role of attention in multisensory integration is not yet clear, a growing amount of evidence highlights their intimate interplay. We addressed how the direction of auditory selective attention modulates brain activity in multisensory integration sites. We used fMRI to measure whole brain BOLD responses while participants heard two auditory speech streams presented simultaneously from the same location but at different pitch. Participants were encouraged to selectively attend one of the two streams, indicated by a pre-trial cue. To promote selection they were tested on memory for words from the cued stream (2AFC) after each trial. Concurrent with the two speech streams, a video-clip of the mouth of a speaker was presented from the centre of the display. The mouth could match the cued audio stream, the uncued one, or neither, with equal probability. The remarkable finding was that, for physically identical conditions, we found a significant BOLD modulation in the Superior Temporal Sulcus (STS) which solely depended on participants' direction of attention. That is, with the two speech streams present, attending to the one congruent with the mouth enabled STS activity, compared to attending the incongruent one. This result demonstrates that selective attention can strongly determine the physiological expression of audiovisual integration through modulation of the activity in associative areas, such as the STS. The behavioural consequence of this effect was confirmed with another experiment, whereby audiovisual enhancement in speech perception was seen only for attended speech streams thus supporting the significant modulatory role of attention in MSI.

Keywords

fMRI, attention, speech, audio-visual

Abstract from the 14th International Multisensory Research Forum, The Hebrew University of Jerusalem, Israel, 2013.

* Contact author: luis.moris.fernandez@gmail.com

DOI:10.1163/22134808-000S0066

BRILL Multisensory Research 26 Supplement (2013) 95 brill.com/msr

Oral Presentation

Visual predictions influence speed and size of auditory responses: An EEG study

Tim Paris *, **Jeesun Kim and Chris Davis**

MARCS Institute, University of Western Sydney, Australia

Abstract

Dynamic stimuli often provide predictive information about upcoming events occurring in a different modality. From a predictive coding framework, such expected events would require less processing than unexpected ones. We tested this idea using neural and behavioural measures of auditory processing in a vision-predicts-sound paradigm.

In the experiment, participants were presented with a visual stimulus followed by either a low or high tone while EEG was recorded from a 256 channel active system. The visual stimulus was either a picture (no prediction), a movie cuing the time of an upcoming auditory event (Event prediction) or a movie cueing the type of tone (Tone prediction). In one third of trials, interspersed throughout the experiment, participants also responded to the tone with a button press.

In both conditions, participants responded faster to the predicted than the unpredicted sounds. ERP results showed that visual Event predictions resulted in significantly smaller P2 ERP amplitude. Conversely, Tone predictions resulted in greater P2 ERP amplitude. The results will be discussed in terms of a predictive coding account.

Keywords

Prediction, audiovisual, ERP, context

Abstract from the 14th International Multisensory Research Forum, The Hebrew University of Jerusalem, Israel, 2013.

* Contact author: t.paris@uws.edu.au

DOI:10.1163/22134808-000S0067

BRILL Multisensory Research 26 Supplement (2013) 96–97 brill.com/msr

Oral Presentation

Multisensory mechanisms for perceptual disambiguation. A classification image study on the stream–bounce illusion

Cesare V. Parise [1,2,*] **and Marc O. Ernst** [2]

[1] Max Planck Institute for Biological Cybernetics, Germany
[2] Department of Cognitive Neuroscience, University of Bielefeld, Germany

Abstract

Sensory information is inherently ambiguous, and a given signal can in principle correspond to infinite states of the world. A primary task for the observer is therefore to disambiguate sensory information and accurately infer the actual state of the world.

Here, we take the stream–bounce illusion as a tool to investigate perceptual disambiguation from a cue-integration perspective, and explore how humans gather and combine sensory information to resolve ambiguity.

In a classification task, we presented two bars moving in opposite directions along the same trajectory meeting at the centre. We asked observers to classify such ambiguous displays as streaming or bouncing. Stimuli were embedded in dynamic audiovisual noise, so that through a reverse correlation analysis, we could estimate the perceptual templates used for the classification. Such templates, the classification images, describe the spatiotemporal statistical properties of the noise, which are selectively associated to either percept. Our results demonstrate that the features of both visual and auditory noise, and interactions thereof, strongly biased the final percept towards streaming or bouncing.

Computationally, participants' performance is explained by a model involving a matching stage, where the perceptual systems cross-correlate the sensory signals with the internal templates; and an integration stage, where matching estimates are linearly combined to determine the final percept. These results demonstrate that observers use analogous MLE-like integration principles for categorical stimulus properties (stream/bounce decisions) as they do for continuous estimates (object size, position, etc.).

Abstract from the 14th International Multisensory Research Forum, The Hebrew University of Jerusalem, Israel, 2013.

* Contact author: cesare.parise@tuebingen.mpg.de

 DOI:10.1163/22134808-000S0068

Finally, the time-course of the classification images reveal that most of the decisional weight for disambiguation is assigned to information gathered before the physical crossing of the stimuli, thus highlighting a predictive nature of perceptual disambiguation.

Keywords
Perceptual ambiguity, stream–bounce illusion, classification image, computational modeling

BRILL Multisensory Research 26 Supplement (2013) 98 brill.com/msr

Oral Presentation

Visual experience and the establishment of spatial and tactile maps in the brain

Michael J. Proulx [1,*], **Achille Pasqualotto** [1] **and Martin I. Sereno** [2,3]

[1] University of Bath, UK

[2] University College London, Birkbeck College University of London, UK

[3] University of California, San Diego, CA, USA

Abstract

Visual experience is necessary for the normal development of spatial cognition. Although some aspects of spatial processing in an egocentric reference frame are well preserved in the case of congenital blindness, it appears the lack of vision influences processing in an allocentric reference frame. Here I will first report our finding that visual experience biases congenitally blind participants to rely on an egocentric reference frame for spatial cognition tasks, rather than an allocentric reference frame that is favoured by late blind and sighted participants. This suggests that visual experience can influence the neural basis of spatial representations, given the importance of vision for normal multisensory processing. A previous study discovered an area in human parietal cortex with aligned maps of tactile and visual stimuli. We investigated whether visual experience is necessary for the development of tactile maps in multisensory areas, such as the putative human ventral intraparietal (VIP) area reported previously. Congenital and late blind participants underwent functional magnetic resonance imaging while exposed to facial air-puffs of varying polar angle. Phase-encoding of the facial air-puffs allowed for tactile mapping that revealed full face maps in the posterior parietal cortex of the late blind participants, as reported previously for sighted participants. The congenitally blind participants however did not have the same representation of the face map as those with visual experience. Retinotopic maps are ideal for characterizing an allocentric representation of the environment and might be necessary as the basis for representing and aligning the neural representation of space as seen in the somatotopic maps examined here.

Keywords

Blindness, spatial cognition, parietal cortex, somatosensory

Abstract from the 14th International Multisensory Research Forum, The Hebrew University of Jerusalem, Israel, 2013.

* Contact author: m.j.proulx@bath.ac.uk

 DOI:10.1163/22134808-000S0069

BRILL Multisensory Research 26 Supplement (2013) 99 brill.com/msr

Oral Presentation

Odours modulate early neural responses to matching visual objects

Amanda Robinson [1,*], **Judith Reinhard** [1] **and Jason Mattingley** [1,2]

[1] Queensland Brain Institute, The University of Queensland, Australia
[2] School of Psychology, The University of Queensland, Australia

Abstract
Sensory information is initially registered within anatomically and functionally segregated brain networks, but is also integrated across modalities in higher cortical areas. Although considerable research has focused on uncovering the neural correlates of multisensory integration for the modalities of vision and audition, much less effort has been devoted to understanding interactions between vision and olfaction. Here we asked whether odours affect neural activity evoked by images of familiar visual objects that carry characteristic smells (e.g., orange, mint, rose). We employed scalp-recorded electroencephalography to measure visual event-related potentials (ERPs) to briefly presented photographs of familiar, odour-related objects, such as an orange or mint leaves. On each trial, participants ($N = 26$) inhaled either a matching odour (e.g., orange scent for a photograph of an orange fruit), a non-matching odour (e.g., mint scent for a photograph of an orange fruit), an irrelevant odour (coffee), or plain air. Participants maintained their attention on the visual stimuli by judging the shape of a border surrounding each picture. The N1 component of the visual ERP, 100–170 ms after image presentation, was reliably enhanced for matching odours in female participants, but not in males. This is consistent with evidence that females are superior in detecting, discriminating and identifying odours, and that they have a higher grey matter concentration in olfactory areas of the orbitofrontal cortex. We conclude that early visual processing is influenced by olfaction due to learned associations between odours and the objects that emit them, and that these associations are stronger in females than males.

Keywords
Olfaction, vision, event-related potentials, electroencephalography

Abstract from the 14th International Multisensory Research Forum, The Hebrew University of Jerusalem, Israel, 2013.
* Contact author: amanda.robinson@uqconnect.edu.au

DOI:10.1163/22134808-000S0070

BRILL Multisensory Research 26 Supplement (2013) 100 brill.com/msr

Oral Presentation

Modulating tactile perception with rhythmic TMS entrainment

Manuela Ruzzoli [1,*] **and Salvador Soto-Faraco** [1,2]

[1] Center for Brain and Cognition (CBC), Universitat Pompeu Fabra, Spain
[2] Institució Catalana de Recerca i Estudis Avançats (ICREA), Spain

Abstract

It has been claimed that oscillatory firing activity plays a functional role in neural systems. Evidence suggests that rhythmic activity in the alpha range (~10 Hz) has a causal effect in visual perception and attention deployment (Romei *et al.*, 2008). The present study addresses the causal role of brain rhythms in tactile processing. We used rhythmic TMS to entrain specific frequencies (10 and 20 Hz) focusing on the parietal lobe (IPS) and primary somatosensory areas (S1). The results revealed that entrainment of alpha frequency in the IPS decreases contralateral somatosensory detection and enhances ipsilateral performance [experiment 1]. Interestingly, we found that this TMS-induced ipsilateral enhancement effect abides with an external frame of reference, not to the anatomical side [experiment 2]. Furthermore, the effect of pre-stimulus entrainment of 10 and 20 Hz activity in S1 is not univocal when arrhythmic TMS is engaged [experiment 3]. Altogether, the present findings suggest that the functional role in tactile processing of alpha frequency in IPS is specific to the interaction with the external world, indicating similarities between senses (vision and touch) in the communication between top-down (parietal) and primary sensory areas (V1 or S1). Pre-stimulus S1 rTMS might affect performance regardless of the frequency engaged, that is, when the stimulus is absent (pre-stimulus) TMS decreases cortical excitability in S1.

Keywords

Tactile processing, rhythmic TMS, intra-parietal sulcus, primary somatosensory cortex

Reference

Romei, V., *et al.* (2008). *Cereb. Cortex* **18**(9), 2010–2018.

Abstract from the 14th International Multisensory Research Forum, The Hebrew University of Jerusalem, Israel, 2013.

* Contact author: manuela.ruzzoli@gmail.com

 DOI:10.1163/22134808-000S0071

BRILL Multisensory Research 26 Supplement (2013) 101 brill.com/msr

Oral Presentation

The effects of reverberation on the spatio-temporal synchrony of moving audio-visual stimuli

Narayan Sankaran *, **Simon Carlile and Johahn Leung**
University of Sydney, Australia

Abstract
The current study investigated the temporal dynamics of audio-visual moving targets, examining the effects of velocity and acoustical environmental cues on the sensitivity of audio-visual cohesion. The acoustical stimuli were generated using Virtual Auditory Space, and two types of moving sounds were presented: (1) an immersive stimulus using Binaural Room Impulse Responses (BRIRs), that were recorded for each subject *in situ* in the testing room; thereby containing the combined spectral and temporal acoustic transformation due to both head-related filtering and the room's characteristic response; (2) an environmentally impoverished stimulus using individualised Head-Related Impulse Responses (HRIRs) recorded in an anechoic chamber that lacked any reverberant cues. Apparent visual motion was generated by the sequential activation of LED's along the same radial path as the auditory motion, 1 m away from the subjects. Moving stimuli were presented with a random temporal offset between each stimuli ranging from [−400 ms, 400 ms] (auditory or visual leading). Using a 2AFC paradigm, subjects reported which modality they perceived as leading for velocities of 25, 50 and 100°/s. At 25°/s the stimuli were perceived to be coincident when the visual stimulus was actually leading but, counter intuitively, the offset decreased as a function of increased velocity. The offset at PSE was greater for HRIRs than for BRIRs but the psychometric functions were significantly steeper for the HRIRs suggesting greater spatio-temporal resolution. This may reflect the degradation of the auditory cues produced by reverberation.

Keywords
Multimodal integration, motion perception, virtual auditory space, environmental acoustics, binaural room impulse response

---

Abstract from the 14th International Multisensory Research Forum, The Hebrew University of Jerusalem, Israel, 2013.
* Contact author: nsan4325@uni.sydney.edu.au

 DOI:10.1163/22134808-000S0072

BRILL Multisensory Research 26 Supplement (2013) 102 brill.com/msr

Oral Presentation

Interaction of embodied cognition and social perception in the primary somatosensory cortex

Michael Schaefer *, **Hans-Jochen Heinze and Michael Rotte**

University of Magdeburg, Germany

Abstract

In the traditional view our brain represents the body in an area called the primary somatosensory cortex. An increasing body of evidence demonstrated that this brain region represents the perceived body image rather than reflecting mechanical stimulation on the body surface. It has been shown that the primary somatosensory cortex plays a role in multisensory integration of visuotactile stimuli, in empathy, and even in social perception. Here we want to further examine the role of the somatosensory cortices in social perception. We present results of an fMRI study showing how incidental touch sensations influence social judgments in a metaphor-specific way. Participants experienced rough or smooth touch before being asked to judge an ambiguous social interaction. Results revealed that rough touch made social interactions appear more difficult and adversarial, consistent with a rough metaphor (compared with smooth touch or no touch). This impact of tactile cues on social impressions was accompanied by a network including primary somatosensory cortex, amygdala, hippocampus and right inferior prefrontal cortex. Hence, receiving touch from a rough object seems to trigger the application of associated ontological concepts (or scaffolds) even for unrelated people and situations (but not to unrelated or more general feelings). Since this priming was based on somatosensory brain areas, our results provide support for the theory that sensorimotor grounding is intrinsic to cognitive processes.

Keywords

Body, somatosensory cortex, social perception

Abstract from the 14th International Multisensory Research Forum, The Hebrew University of Jerusalem, Israel, 2013.

* Contact author: mischa@neuro2.med.uni-magdeburg.de

 DOI:10.1163/22134808-000S0073

BRILL

Multisensory Research 26 Supplement (2013) 103

MULTISENSORY RESEARCH

brill.com/msr

Oral Presentation

Alternating speed on dance videos influences duration judgments in dancers and non-dancers

Helena Sgouramani [1,*] and Argiro Vatakis [1,2]

[1] Postgraduate Program Basic and Applied Cognitive Science, Department of Philosophy and History of Science, University of Athens, Greece

[2] Cognitive Systems Research Institute (CSRI), Athens, Greece

---

Abstract

Speed has been proposed as a modulating factor on duration estimation. However, the different measurement methodologies and experimental designs used have led to inconsistent results across studies, and, thus, the issue of how speed modulates time estimation remains unresolved. Additionally, no studies have looked into the role of spatiotemporal experience and susceptibility to modulations of speed in timing judgments. In the present study, therefore, using naturalistic, dynamic dance stimuli, we aimed at defining the role of speed and the interaction of speed and experience on time estimation. We presented videos of a dancer performing identical ballet steps in fast and slow versions, while controlling for the number of changes present. Professional dancers and non-dancers performed duration judgments through a production and reproduction task. Analysis revealed a significantly larger underestimation of fast videos as compared to slow ones during reproduction. The exact opposite result was true for the production task. Dancers were generally more accurate and significantly less variable in their time estimations as compared to non-dancers. Speed and experience, therefore, affect participants' estimates of time. Results are discussed in association to the theoretical framework of current models by focusing on the role of attention.

Keywords

Time estimation, reproduction task, production task, speed, dance, attention

---

Abstract from the 14th International Multisensory Research Forum, The Hebrew University of Jerusalem, Israel, 2013.

* Contact author: elenisgou@live.com

DOI:10.1163/22134808-000S0074

BRILL Multisensory Research 26 Supplement (2013) 104–105 brill.com/msr

Oral Presentation

Minimum audible angle at the subjective front during listener's head rotation

Yôiti Suzuki [1,*], **Kagesho Ohba** [1], **Shuichi Sakamoto** [1], **Akio Honda** [2] and **Yukio Iwaya** [3]

[1] Research Institute of Electrical Communication and Graduate School of Information Sciences, Tohoku University, Japan

[2] Department of Welfare Psychology, Tohoku Fukushi University, Japan

[3] Faculty of Engineering, Tohoku Gakuin University, Japan

Abstract

Sound localization is a multisensory process consisting not only of hearing, but also of self-motion perception. A large number of studies have shown that the listener's head movement, particularly horizontal rotation, effectively improves sound localization acuity (Wallach, 1939; Thurlow, 1967; Kawaura, 1989; etc.). However, research into sound localization during head rotation is scarce. Thus, the multisensory process underlying sound localization remains unclear.

In the present study, we directly investigated the minimum audible angle (MAA) — the detection threshold needed for the listener to localize sounds at different positions — at the front during horizontal head rotation. A sound stimulus (30-ms noise burst) was presented from a loudspeaker of a circular array (r = 1.1 m), with a loudspeaker separation of 2.5 degrees. The listener, sitting at the center of the circle, was asked to answer whether the sound stimulus was presented from the left or right of the subjective front (2AFC). We considered two listening conditions, static and dynamic. In the static condition, listeners were asked to keep their heads still. Meanwhile, for the dynamic condition, listeners were asked to rapidly rotate their heads; the test stimulus was triggered during head movement. Sound stimuli were presented from a randomly selected loudspeaker out of a subset of 13 centered at the listener's physical front at the time of stimulus presentation. Results showed the MAA to deteriorate significantly in the dynamic condition. This implies that the improvements in sound localization due to head motion could be explained by the multiple-look model (Viemeister and Wakefield, 1991).

Abstract from the 14th International Multisensory Research Forum, The Hebrew University of Jerusalem, Israel, 2013.

* Contact author: yoh@riec.tohoku.ac.jp

 DOI:10.1163/22134808-000S0075

Keywords
Active listening, sound localization, self motion, head rotation

References

Kawaura, J., *et al.* (1989). Sound localization in headphone reproduction by simulating transfer functions from the sound source to the external ear, *J. Acoust. Soc. Jpn.* **45**(10), 756–766 (in Japanese).

Thurlow, W. R. and Runge, P. S. (1967). Effect of induced head movement in localization of direction of sound, *J. Acoust. Soc. Am.* **42**, 480.

Viemeister, N. F. and Wakefield, G. H. (1991). Temporal integration and multiple looks, *J. Acoust. Soc. Am.* **90**, 858–865.

Wallach, H. (1939). On sound localization, *J. Acoust. Soc. Am.* **10**, 270.

BRILL

Multisensory Research 26 Supplement (2013) 106

MULTISENSORY RESEARCH

brill.com/msr

Oral Presentation

Summation of duration within and across senses

Kohske Takahashi * **and Katsumi Watanabe**

The University of Tokyo, Japan

Abstract

It has been shown that sensory-specific or common processes mediate perception and memory. Here we investigated whether the summation operation of temporal duration is specific to sensory modalities. Two filled intervals were presented as intra- or inter-modal sequence (audition–audition; vision–vision; audition–vision; vision–audition). Participants reproduced the sum (0.6–1.1 s) of two intervals. We found that the summation operation of two intervals generally led to an overestimation of the sum, irrespective of conditions. The variations of reproduction were comparable between inter- and intra-modal conditions, indicating no impairment in inter-modal summation of duration. Furthermore, the variations showed positive within-participant correlation between all modality conditions (rs > 0.74). These results suggested that a supra-modal process might mediate the summation of duration. In the subsequent experiment, the similar pattern of results was observed by using short intervals (to be summed to 0.3–0.5 s); however, the intra-modal conditions showed much reduced correlations with other conditions ($r = 0.39$), which indicated that the summation of short durations might be mediated by sensory-specific processes. In sum, the present results suggest two processes for summation of temporal durations: the sensory-specific process that operates only on short durations, and the other supra-modal process that handles otherwise (i.e., when longer durations are to be summed or when inter-modal summation is required).

Keywords

Time perception, duration, reproduction

Abstract from the 14th International Multisensory Research Forum, The Hebrew University of Jerusalem, Israel, 2013.

* Contact author: takahashi.kohske@gmail.com

DOI:10.1163/22134808-000S0076

BRILL
Multisensory Research 26 Supplement (2013) 107
brill.com/msr

Oral Presentation

Audiovisual onset differences are used to determine the identity of ambiguous syllables

Sanne ten Oever [1,*], **Alexander Sack** [2], **Katherine L. Wheat** [2], **Nina Bien** [2,3] **and Nienke van Atteveldt** [2,4]

[1] Department of Cognitive Neuroscience, University Maastricht, The Netherlands
[2] Maastricht University, The Netherlands
[3] University of Luxembourg, The Netherlands
[4] Netherlands Institute for Neuroscience, The Netherlands

Abstract

Content and temporal cues have been shown to interact during audiovisual (AV) speech identification. Typically, the most reliable unimodal cue is used to identify specific speech features; however, visual cues are only used if the audiovisual stimuli are presented within a certain temporal integration window (TWI). This suggests that temporal cues denote whether unimodal stimuli belong together and should be integrated. It is unknown whether temporal cues also provide information about speech content. Since spoken syllables have naturally varying audiovisual onset asynchronies, we hypothesize that for suboptimal AV cues presented within the TWI, these natural AV onset differences can aid in speech identification. To test this, we presented low-intensity auditory syllables concurrently with visual speech signals, and varied the stimulus onset asynchronies (SOA) of the audiovisual pair, while participants identified the syllables. We revealed that the most reliable cues of the audiovisual input were used to identify specific speech features (e.g., voicing). Additionally, we showed that the TWI was wider for congruent stimulus pairs. Finally, we found a specific response pattern across the SOA-range for syllables that were not reliably identified by the unimodal cues, which we explained by the use of natural onset differences between audiovisual speech signals. This indicates that temporal cues not only determine whether or not different inputs belong together, but additionally convey identity information of audiovisual pairs. These results provide a detailed behavioral basis for further neuro-imaging and stimulation studies to unravel the neurofunctional mechanisms of the audiovisual–temporal interplay within speech perception.

Keywords

Stimulus onset asynchrony, audiovisual speech, temporal relationship

---

Abstract from the 14th International Multisensory Research Forum, The Hebrew University of Jerusalem, Israel, 2013.

* Contact author: sanne.tenoever@maastrichtuniversity.nl

DOI:10.1163/22134808-000S0077

BRILL Multisensory Research 26 Supplement (2013) 108 brill.com/msr

Oral Presentation

Catch me if you can — perceived distance in a function of action

Oliver M. Toskovic *

Laboratory of Experimental Psychology, Faculty of Philosophy, Belgrade, Serbia

Abstract

Distances toward zenith are perceived as longer than physically equal distances towards horizon. Reason for this anisotropy might be in better action performance, since, in order to reach something towards zenith, we oppose gravity, and if perceived distance is longer, we would put more effort and easily oppose gravity. If this is true, action towards the ground would be in line with gravity, and perceived distance would be shorter. We tested this hypothesis in two experiments in which 27 participants (14 + 13) had the task to equalize the perceived distances of stimuli on three directions (horizontal, tilted 45 degrees and vertical). In the first experiment tilted and vertical directions were toward zenith, and in the second toward the ground. One of the stimuli was the standard, and participants matched distances of other two with the standard. Results of the first experiment show significant difference in matched distances between 0 and 45 degrees ($F(1, 13) = 67.17$, $p < 0.01$), 0 and 90 degrees ($F(1, 13) = 122.76$, $p < 0.01$), 45 and 90 degrees ($F(1, 13) = 41.49$, $p < 0.01$). For the second experiment, results show significant difference between 0 and 45 degrees ($F(1, 12) = 12.85$, $p < 0.01$), 0 and 90 degrees ($F(1, 12) = 29.89$, $p < 0.01$), and no significant difference between 45 and 90 degrees direction. Participants matched longer horizontal distances with shorter distances toward zenith, meaning that perceived distances toward zenith were longer. On the other hand, they matched shorter horizontal distances with longer distances towards the ground, meaning that perceived distances towards the ground were shorter. Both of these findings are in line with our hypothesis.

Keywords

Anisotropy, perceived distance, gravity, perception and action, viewing direction

Abstract from the 14th International Multisensory Research Forum, The Hebrew University of Jerusalem, Israel, 2013.

* Contact author: otoskovi@f.bg.ac.rs

 DOI:10.1163/22134808-000S0078

BRILL Multisensory Research 26 Supplement (2013) 109 brill.com/msr

Oral Presentation

Contextual control of audiovisual integration in low-level sensory cortices

Nienke M. van Atteveldt [1,2,*], **Bradley S. Peterson** [3] **and Charles E. Schroeder** [3,4]

[1] Netherlands Institute for Neuroscience, Amsterdam, The Netherlands
[2] Maastricht University, The Netherlands
[3] Columbia University, Dept. Psychiatry, New York, NY, USA
[4] Nathan S. Kline Institute, Cognitive Neuroscience & Schizophrenia Program, Orangeburg, NY, USA

Abstract

Potential sources of multisensory influences on low-level sensory cortices include direct projections from sensory cortices of different modalities, as well as more indirect feedback inputs from higher order multisensory cortical regions. These multiple architectures may be functionally complementary, but the exact roles and inter-relationships of the circuits are unknown. We tested the propositions that: (1) Feedforward and lateral pathways subserve speed functions, such as fast detection of potentially dangerous objects approaching in the periphery. Multisensory integration effects in this context are predicted in peripheral fields of low-level sensory cortices. (2) Slower feedback pathways underpin accuracy functions, such as discrimination or identification of complex sensory inputs. Integration effects in this context are predicted in higher-order association cortices and central/foveal fields of low-level sensory cortex. We used functional magnetic resonance imaging to compare the effects of central *versus* peripheral stimulation on audiovisual integration, while varying speed and accuracy requirements for behavioral responses. We found that task demands and stimulus eccentricity interact in low-level cortices in more complex ways than predicted by the simple dichotomy of peripheral/speed and foveal/accuracy functions. Interestingly, different task requirements yielded seemingly qualitatively different integration processes in visual *versus* auditory cortex. Additionally, our findings point to individual differences in integration that may be related to skills and strategy. Overall, these findings suggest that instead of using fixed, specialized pathways, the exact circuits and mechanisms that are used for low-level multisensory integration are much more flexible and contingent upon both individual and contextual factors than previously assumed.

Keywords

Audiovisual, low-level integration, contextual control, fMRI, task demand, eccentricity

---

Abstract from the 14th International Multisensory Research Forum, The Hebrew University of Jerusalem, Israel, 2013.

* Contact author: n.vanatteveldt@maastrichtuniversity.nl

 DOI:10.1163/22134808-000S0079

BRILL Multisensory Research 26 Supplement (2013) 110 brill.com/msr

Oral Presentation

Subcortical, modality-specific pathways contribute to multisensory processing in humans

Rudy van den Brink [1,*], Mike X. Cohen [2], Erik van der Burg [3], Durk Talsma [4], Vissers Marlies [2] and Heleen A. Slagter [2]

[1] Leiden University, The Netherlands
[2] University of Amsterdam, The Netherlands
[3] The University of Sydney, Australia
[4] Ghent University, Belgium

Abstract

Oftentimes, we perceive our environment by integrating information across multiple senses. Recent studies suggest that such integration occurs at much earlier processing stages than once thought possible, including in thalamic nuclei and putatively unisensory cortical brain regions. Here, we used diffusion tensor imaging (DTI) and an audiovisual integration task to test the hypothesis that anatomical connections between sensory-related subcortical structures and sensory cortical areas govern multisensory processing in humans. Twenty-five subjects (mean age 22 years, 22 females) participated in the study. In line with our hypothesis, we show that estimated strength of white-matter connections between the first relay station in the auditory processing stream (the cochlear nucleus), the auditory thalamus, and primary auditory cortex predicted one's ability to combine auditory and visual information in a visual search task. This finding supports a growing body of work that indicates that subcortical sensory pathways do not only feed forward unisensory information to the cortex, and suggests that anatomical brain connectivity contributes to multisensory processing ability in humans.

Keywords

Brain, diffusion tensor imaging, multisensory, multisensory integration, structural connectivity

Abstract from the 14th International Multisensory Research Forum, The Hebrew University of Jerusalem, Israel, 2013.

* Contact author: rudyvdbrink@gmail.com

DOI:10.1163/22134808-000S0080

BRILL Multisensory Research 26 Supplement (2013) 111 brill.com/msr

Oral Presentation

Proprioception can contribute to the localization of visual targets for goal-directed movements

Marieke C. W. van der Graaff *

MOVE Research Institute Amsterdam, Faculty of Human Movement Sciences, VU University Amsterdam, The Netherlands

Abstract

When we want to make accurate hand movements to for instance a visual target, it is assumed that we combine visual and proprioceptive information to localize our hand precisely. It has been suggested that you do the same for visual targets. In this case, visual localization is based on the target's position relative to your fovea, and proprioceptive localization is relative to your hand. If you cannot see your hand, this proprioceptive target information will degrade, as updating the location of the target relative to your moving hand adds noise. If this proprioceptive estimate of the target is relevant, movements will become more precise and accurate if we are able to keep a proprioceptive estimate, for instance by placing the other (invisible) hand near the target. This unseen hand does not add new information; it only copies the visual information to proprioception. We tested this prediction by asking subjects to repeatedly move their right index finger between several visual targets on a table. We repeated the experiment after the subjects put their left index finger under one of the target locations (without any feedback). Subjects made considerable errors in positioning their invisible right finger. As predicted, adding a non-informative finger improved the accuracy of the movements (by about 20%). The bias of the invisible right hand after repetitive movements was on average shifted towards the left hand's location. Our results support the hypothesis that for goal-directed movements, we use proprioceptive estimates of visual targets.

Keywords

Proprioception, vision, goal-directed movements

Abstract from the 14th International Multisensory Research Forum, The Hebrew University of Jerusalem, Israel, 2013.

* Contact author: m.c.w.vander.Graaff@vu.nl

 DOI:10.1163/22134808-000S0081

BRILL Multisensory Research 26 Supplement (2013) 112 brill.com/msr

Oral Presentation

Dissociating pitch and loudness interactions between audition and touch

Jeffrey Yau *

Johns Hopkins University, Baltimore, MD, USA

Abstract

In this talk I review a series of studies focused on characterizing the relationship between audition and touch. We perceive the frequency and intensity of environmental oscillations (sounds and vibrations) using both modalities. Audio-tactile interactions in the frequency domain are frequency-specific and bi-directional: Interaction patterns support the existence of shared (supramodal) frequency representations. In contrast, audio-tactile interactions in the intensity domain reveal a separate set of integration rules. Thus, a pair of auditory and tactile inputs combines differently depending on the perceptual task (i.e., pitch *vs.* loudness discrimination). That distinct rules govern the integration of auditory and tactile signals in pitch and loudness perception implies that the two processes rely on separate neural mechanisms. Other perceptual processes combining auditory and tactile signals, like stimulus detection or spatial localization, may also adhere to unique integration rules that reflect dissociable neural mechanisms. These results underscore the complexity and specificity of multisensory interactions.

Keywords

Psychophysics, integration, attention

Abstract from the 14th International Multisensory Research Forum, The Hebrew University of Jerusalem, Israel, 2013.

* Contact author: jeff.yau@gmail.com

DOI:10.1163/22134808-000S0082

BRILL Multisensory Research 26 Supplement (2013) 113–114 brill.com/msr

Oral Presentation

Supramodal processing in visual learning and plasticity

Nicolas Zilber [1,2,3,*], **Philippe Ciuciu** [1,2,3], **Alexandre Gramfort** [1,2,3,4] and **Virginie van Wassenhove** [1,5,6]

[1] CEA, DSV/I2BM, NeuroSpin Center, Gif-sur-Yvette, France
[2] INRIA, Parietal team, Saclay, France
[3] LNAO/NeuroSpin, CEA Saclay, Gif-sur-Yvette, France
[4] Institut Mines-Télécom, Télécom ParisTech, CNRS LTCI, Paris, France
[5] INSERM, U992, Cognitive Neuroimaging Unit, Gif-sur-Yvette, France
[6] Univ. Paris-Sud, Cognitive Neuroimaging Unit, Gif-sur-Yvette, France

Abstract

Multisensory interactions are ubiquitous in cortex and question whether sensory cortices can be distinctively supramodal i.e., capable of functional selectivity irrespective of the sensory modality of inputs (Pascual-Leone and Hamilton, 2001; Ricciardi and Pietrini, 2011; Voss and Zatorre, 2012). This suggests that visual perceptual learning could benefit from supramodal processing *via* reverse hierarchy (Ahissar and Hochstein, 2004; Proulx *et al.*, 2012). To test this, novel stimuli were developed consisting of acoustic textures sharing the temporal statistics of visual random dot kinematograms (RDKs). Two groups of participants were trained in a difficult visual coherence discrimination task, with or without sounds, while being recorded with magnetoencephalography (MEG). Participants trained in audiovisual conditions (AV) significantly outperformed visual trainees (V) although they were unaware of their progress. When contrasting post- *vs.* pre-training MEG data, significant differences in the dynamic pattern and the cortical regions responsive to visual RDKs were observed between the two groups in response to visual RDKs. Specifically, neural activity in multisensory cortices (mSTS) correlated with post-training performances and visual motion area (hMT+) selectively responded to trained coherence levels but only in the AV trainees. The latencies of these effects suggest selective feedback from mSTS to hMT+ possibly mediated by posterior temporal cortices (pSTS). Altogether, we interpret our results in the context of the Reverse Hierarchy Theory of learning (Ahissar and Hochstein, 2004) in which supramodal processing optimizes visual perceptual learning by capitalizing on sensory-invariant representations namely, global coherence levels across sensory modalities.

Abstract from the 14th International Multisensory Research Forum, The Hebrew University of Jerusalem, Israel, 2013.

* Contact author: nicolaszilber@hotmail.com

 DOI:10.1163/22134808-000S0083

Keywords
Supramodal processing, learning, plasticity, MEG

References

Ahissar, M. and Hochstein, S. (2004). The reverse hierarchy theory of visual perceptual learning, *Trends Cogn. Sci.* **8**, 457–464.

Pascual-Leone, A. and Hamilton, R. (2001). The metamodal organization of the brain, *Prog. Brain Res.* **134**, 427–445.

Proulx, M., Brown, D., Pasqualotto, A. and Meijer, P. (2012). Multisensory perceptual learning and sensory substitution, *Neurosci. Biobehav. Rev.*, in press.

Ricciardi, E. and Pietrini, P. (2011). New light from the dark: what blindness can teach us about brain function, *Curr. Opin. Neurol.* **24**, 357–363.

Voss, P. and Zatorre, R. (2012). Organization and reorganization of sensory-deprived cortex, *Curr. Biol.* **22**, R168–R173.

Section III: POSTER PRESENTATIONS

BRILL Multisensory Research 26 Supplement (2013) 116 brill.com/msr

Poster Presentation

EyeMusic: A colorful experience for the blind

Sami Abboud [1,*], **Shlomi Hanassy** [1], **Shelly Levy-Tzedek** [1,2], **Dina Tauber** [3], **Shachar Maidenbaum** [1] and **Amir Amedi** [1,2]

[1] Department of Medical Neurobiology, The Institute for Medical Research Israel–Canada, Faculty of Medicine, The Hebrew University of Jerusalem, Israel

[2] The Edmond and Lily Safra Center for Brain Sciences (ELSC), The Hebrew University of Jerusalem, Israel

[3] Faculty of Medicine, The Hebrew University of Jerusalem, Israel

Abstract

Various sensory substitution devices (SSDs) have been designed in the past decades with the aim of creating a visual experience for the blind. We will present the EyeMusic, a visual-to-auditory SSD for the blind, providing both shape and color information. Our design uses musical notes generated by natural instruments to convey the visual information. This creates a pleasant user experience and facilitates prolonged use. We further present behavioral results showing that shape and color information are indeed decodable from the generated auditory stimuli.

Keywords

Sensory-substitution devices (SSD), blindness, color vision

References

Levy-Tzedek, S., Novick, I., Arbel, R., Abboud, S., Maidenbaum, S., Vaadia, E. and Amedi, A. (2012). Cross-sensory transfer of sensory-motor information: visuomotor learning affects performance on an audiomotor task, using sensory-substitution, *Scientific Reports* **2**, 949.

Levy-Tzedek, S., Hanassy, S., Abboud, S., Maidenbaum, S. and Amedi, A. (2012). Fast, accurate reaching movements with a visual-to-auditory sensory substitution device, *Restorative Neurology and Neuroscience* **30**(4), 313–323.

Reich, L., Szwed, M., Cohen, L. and Amedi, A. (2012). A ventral visual stream reading center independent of visual experience, *Current Biology* **21**, 1–6.

Striem-Amit, E., Dehaene, S., Cohen, L. and Amedi, A. (2012). Reading with sounds: Sensory substitution selectively activates the visual word form area in the blind, *Neuron*.

Abstract from the 14th International Multisensory Research Forum, The Hebrew University of Jerusalem, Israel, 2013.

* Contact author: sami.abboud@mail.huji.ac.il

 DOI:10.1163/22134808-000S0084

BRILL Multisensory Research 26 Supplement (2013) 117 brill.com/msr

Poster Presentation

Perception of rhythm through auditory, vibro-tactile and visual stimulations: An fMRI study

Rodrigo Araneda *, **Elodie Lerens, Laurent Renier and Anne De Volder**

Université Catholique de Louvain, Belgium

Abstract

Recent studies reported an involvement of motor/premotor brain areas during the perception of rhythm and beat in audition. However, little is known about the neural network of beat perception through non-auditory modalities. Here we used functional magnetic resonance imaging with auditory, vibro-tactile and visual rhythmic sequences in order to highlight the modality-specific areas involved in beat perception. We contrasted the brain activity changes in 20 healthy volunteers exposed to rhythmic sequences with a beat and control sequences without a beat. Results showed a recruitment of premotor cortex, supplementary motor area (SMA) and basal ganglia during beat sequences compared to rest in all three modalities. Beat sequences compared to no-beat sequences elicited the selective recruitment of the putamen and the SMA in the three sensory modalities, although to a lesser extent in vision. We conclude that the putamen and the SMA play a role in the prediction ability of a regular temporal pattern, although this seems less pronounced in vision compared to audition and to the vibro-tactile sensory modality.

Abstract from the 14th International Multisensory Research Forum, The Hebrew University of Jerusalem, Israel, 2013.

* Contact author: araneda.rodrigoantonio@gmail.com

 DOI:10.1163/22134808-000S0085

BRILL Multisensory Research 26 Supplement (2013) 118 brill.com/msr

Poster Presentation

Perceptual ambiguity — perception and processing of spatially discordant/concordant audiovisual stimuli

Jan Bennemann [1,*], Philipp Benner [2], Claudia Freigang [1], Wiktor Mlynarski [2], Marc Stöhr [1] and Rudolf Rübsamen [1]

[1] Leipzig University, Germany
[2] Max Planck Institute for Mathematics in the Sciences, Germany

Abstract

It is well known that under certain conditions crossmodal interactions alter our perception of multisensory events, especially in conflicting/ambiguous situations, e.g., presenting physically separated (incongruent) audiovisual cues. In case of the ventriloquist illusion, subjects shift the location of an acoustic signal to a spatially discordant visual cue (perceptual fusion), i.e., subjects perceptually unify visual and acoustic stimuli as coming from the same source although they are spatially separated.

In a behavioral experiment applying a simple unification task we presented audiovisual stimuli in the freefield either spatially discordant or spatially concordant. Participants were required to judge spatial relation of co-occurring events (separation/unity). The focus of the present study was to quantify participants' response behavior using a recently developed nonparametric adaptive sequential sampling procedure (Poppe *et al.*, 2012). We tested four acoustic reference positions in central and para-central space (±8° and ±25° corresponding to left (−) and right (+) hemispace). Deviating visual signals were presented between 30° left and right of acoustic signals. Acquired psychometric functions allowed a detailed analysis of individual response behavior across all sampled conditions. Additionally, estimations of 50% correct response thresholds were used as guideline for the parameter setting of audiovisual stimuli in subsequent EEG experiments. We tested the hypothesis according to which the prestimulus cortical activity modulates the perceptual organization of a given stimulus and therefore determines the perceptual outcome of ambiguous audiovisual stimuli. For that EEG was recorded while presenting physically identical stimuli which, however, generate a perception that varies between two possible alternatives.

Keywords

Audiovisual, ambiguity, spatial processing, EEG

Reference

Poppe, S., Benner, P. and Elze, T. (2012). A predictive approach to nonparametric inference for adaptive sequential sampling of psychophysical experiments, *J. Math. Psychol.* **56**(3), 179–195.

Abstract from the 14th International Multisensory Research Forum, The Hebrew University of Jerusalem, Israel, 2013.

* Contact author: jan.bennemann@uni-leipzig.de

 DOI:10.1163/22134808-000S0086

BRILL Multisensory Research 26 Supplement (2013) 119–120 brill.com/msr

Poster Presentation

The feeling of vision: Induction of mirror-touch synaesthesia by increasing of somatosensory cortical excitability

Nadia Bolognini [1,*], **Carlo Miniussi** [2], **Selene Gallo** [1] **and Giuseppe Vallar** [1]

[1] University of Milano Bicocca, Italy
[2] University of Brescia, Italy

Abstract

Brain imaging studies have shown the existence of a mirror network for touch, which allows for an automatic and unconscious simulation of others' somatic states (Keysers *et al.*, 2010). In everyday life, we are typically unaware of this process likely because the system is physiologically active below the threshold of perceptual awareness. However, in persons with mirror-touch synaesthesia, the sight of a touch on another person elicits conscious tactile experiences on their own bodies (Blakemore *et al.*, 2005). In a sham-controlled study (healthy participants, $N = 32$), by combining anodal transcranial direct current stimulation (tDCS, 1.5 mA, 16 min) with a vision-touch interference task developed for studying mirror-touch synaesthesia (Banissy and Ward, 2007), we show that the enhancement of cortical excitability in the primary somatosensory cortex promotes the emergence of synaesthesia-like effects even in non-synaesthetes. Interestingly, these effects are linked with inter-individual differences in cognitive aspects of empathy. Conversely, increasing excitability in the premotor cortex facilitates the integration of spatially congruent viewed and felt touches, without inducing synaesthesia-like effects; the crossmodal facilitation by premotor tDCS is not associated with empathy, thereby confirming the functional dissociation between somatosensory and premotor areas within the tactile mirror system. This evidence indicates that mirror-touch synaesthesia reflects general crossmodal mechanisms associated with emphatic abilities: by increasing the excitability of the tactile mirror system, synaesthesia-like responses may be revealed in non-synaesthetic individuals.

Keywords

Mirror-touch synaesthesia, tDCS

Abstract from the 14th International Multisensory Research Forum, The Hebrew University of Jerusalem, Israel, 2013.

* Contact author: nadia.bolognini@unimib.it

 DOI:10.1163/22134808-000S0087

References

Banissy, M. J. and Ward, J. (2007). Mirror-touch synesthesia is linked with empathy, *Nat. Neurosci.* **10**, 815–816.

Blakemore, S. J., Bristow, D., Bird, G., Frith, C. and Ward, J. (2005). Somatosensory activations during the observation of touch and a case of vision.

BRILL

Multisensory Research 26 Supplement (2013) 121–122

brill.com/msr

Poster Presentation

Visual and auditory spatial signals in naturalistic environments: A computationally-based analysis of functional imaging data

Cecile Bordier *, **Akitoshi Ogawa and Emiliano Macaluso**

Santa Lucia Foundation, Italy

Abstract

A major challenge for fMRI studies is to use realistic stimuli, which is relevant for the understanding of multisensory interactions in the real world. Computational models (Itti *et al.*, 1998; Kayser *et al.*, 2005) have been used to track brain activity during viewing of complex audio-visual stimuli (Bartels *et al.*, 2007; Bordier *et al.*, 2013). We extended this approach to investigate activity associated with high-order spatial aspects in both vision and audition. We utilized a 3D-surround movie that included visual disparity cues and multiple sound sources (centre, front left/right, back left/right). For each visual frame, we computed a disparity map (Liu *et al.*, 2011) and indexed absolute disparity (sum over the entire map) and gradient disparity (local contrast, Bordier *et al.*, 2013). For audition, we indexed sound-spatiality by computing the correlation between each of the 5 external channels and a sixth channel that was delivered over headphones during fMRI (this contained primarily the 'centre' sound). We also indexed the sound intensity contrast (Bordier *et al.*, 2013) to control for mere intensity changes over time. These indexes were used to fit the BOLD signal in 16 subjects, who watched the 3D-surround movie during fMRI. The results showed a dissociation between absolute disparity (PPC and V3A) and gradient disparity (V6, STS and IFG). The auditory–spatiality index correlated with activity in auditory cortex, even after accounting for changes of auditory-intensity. We conclude that the combination of computational models and fMRI enables studying brain activity associated with complex spatial signals in naturalistic multisensory environments.

Keywords

Computational models, visual disparity, multiple sound sources

---

Abstract from the 14th International Multisensory Research Forum, The Hebrew University of Jerusalem, Israel, 2013.

* Contact author: c.bordier@hsantalucia.it

DOI:10.1163/22134808-000S0088

References

Bartels, A., Zeki, S. and Logothetis, N. K. (2007). Natural vision reveals regional specialization to local motion and to contrast invariant, global flow in the human brain, *Cerebral Cortex* **18**, 105–117.

Bordier, C., Puja, F. and Macaluso, E. (2013). Sensory processing during viewing of cinematographic material: computational modeling and functional neuroimaging, *NeuroImage* **67**, 213–226.

Itti, L., Koch, C. and Niebur, E. (1998). A model of saliency-based visual-attention for rapid scene analysis, *IEEE Trans. on Pattern Analysis and Machine Intelligence* **20**, 1254–1259.

Kayser, C., Petkov, C. I., Lippert, M. and Logothetis, N. K. (2005). Mechanisms for allocating auditory attention: an auditory saliency map, *Current Biology* **15**(21), 1943–1947.

Liu, C., Yuen, J. and Torralba, A. (2011). SIFT flow: dense correspondence across different scenes and its applications, *IEEE Trans. Pattern Analysis and Machine Intelligence* **33**(5), 978–994.

BRILL Multisensory Research 26 Supplement (2013) 123 brill.com/msr

Poster Presentation

'Bouba' and 'Kiki' in Namibia? A remote culture make similar shape–sound matches, but different shape–taste matches to westerners

Andrew J. Bremner [1,*], **Serge Caparos** [2], **Jules Davidoff** [1], **Jan W. de Fockert** [1], **Karina Linnell** [1] **and Charles Spence** [3]

[1] Goldsmiths, University of London, UK
[2] Université du Québec à Trois-Rivières, Canada
[3] University of Oxford, UK

Abstract

Western participants consistently match certain shapes with particular speech sounds, tastes, and flavours. Here we demonstrate that the 'Bouba–Kiki effect', a well-known shape–sound symbolism effect commonly observed in Western participants, is also observable in the Himba of Northern Namibia, a remote population with little exposure to Western cultural and environmental influences, and who do not use a written language. However, in contrast to Westerners, the Himba did not map carbonation (in a sample of sparkling water) onto an angular (as opposed to a rounded) shape. Furthermore, they also tended to match less bitter (i.e., milk) chocolate samples to angular rather than rounded shapes; the opposite mapping to that shown by Westerners. Together, these results show that cultural–environmental as well as phylogenetic factors play a central role in shaping our repertoire of crossmodal correspondences.

Keywords

Crossmodal correspondences, crosscultural, sound symbolism, taste, flavour

Abstract from the 14th International Multisensory Research Forum, The Hebrew University of Jerusalem, Israel, 2013.

* Contact author: a.bremner@gold.ac.uk

DOI:10.1163/22134808-000S0089

BRILL Multisensory Research 26 Supplement (2013) 124 brill.com/msr

Poster Presentation

What the temporal dynamics of unimodal sensory estimation can tell us about statistically optimal multimodal integration

Patrick A. Byrne * **and Laurence R. Harris**

Centre for Vision Research, York University, Canada

Abstract

The brain combines noisy, redundant stimulus estimates (e.g., position of an object provided by vision and haptics) in proportion to their reliabilities. The mechanisms used by the brain to assess reliability (inverse of noise) of its estimates are, however, not fully understood. Most models make the untested assumption that internal estimates of a property respond rapidly to fluctuations arising from noise. We tested this assumption by employing a visual stimulus consisting of an array of flickering vertical lines distributed to produce an area of higher density which the subject had to track. This stimulus was similar to one that has been shown (Byrne and Henriques, 2013) to combine optimally with haptic estimates of position. The area of higher density was inconspicuously jumped left or right at varying time intervals before the subject indicated its location relative to a reference line. By looking for whether the jump affected subjects' judgements, we could assess the time required for updating internal position estimates. Subjects' responses showed that large jumps had little effect on position estimates until the post-jump stimulus was present for over 300 ms. At this time the estimate suddenly transitioned to a new location. Updating for smaller jumps proceeded more smoothly. These results imply that position estimates are maintained in an 'attractor', which takes time to shift when sensory information changes. This is inconsistent with current models of how reliability is computed by the brain because such models, which rely on population coding or temporal sampling, require estimates to fluctuate in lock-step with noise.

Keywords

Bayesian, cue-combination, vision, haptics, psychophysics

Reference

Byrne, P. A. and Henriques, D. Y. P. (2012). When more is less: Increasing allocentric visual information can switch visual–proprioceptive combination from an optimal to sub-optimal process, *Neuropsychologia* **51**, 26–37.

Abstract from the 14th International Multisensory Research Forum, The Hebrew University of Jerusalem, Israel, 2013.

* Contact author: dr.patrick.byrne@gmail.com

 DOI:10.1163/22134808-000S0090

BRILL Multisensory Research 26 Supplement (2013) 125 brill.com/msr

Poster Presentation

The contribution of sound in determining the perceptual upright

Michael J. Carnevale * **and Laurence R. Harris**

Centre for Vision Research, York University, Canada

Abstract

The perceived direction of up depends on visual, gravity, and body cues, each of which is given a weighting by the brain (Dyde *et al.*, 2006). Little work has been done, however, to demonstrate whether sound might also be usable by the brain as a cue to up. Here we assess the possible contribution of sound to perceived orientation by adding a sound cue to gravity. The perceptual upright, the direction in which a character is most easily recognized, was assessed using the Oriented Character Recognition Test (OCHART). Subjects identified the character 'p' that was presented in various orientations (0–360 degrees rotation) as either a 'p' or 'd'. The orientations were chosen by a QUEST adaptive staircase procedure and the mean of the points of subjective equality was taken as the perceptual upright. Subjects lay on their side and viewed a laptop screen through a shroud. Thus, body and gravity cues were orthogonal and swings of the perceptual upright towards or away from gravity could be measured. Loudspeakers were mounted above and below the lying subject (opposite the left and right ears) and sounds were presented synchronized to the appearance of the character on the screen. Changes in the direction of the PU were recorded depending on whether a sound was present or not. We conclude that sounds can contribute to the perception of upright.

Keywords

Body orientation, audition, perceptual upright

Reference

Dyde, R. T., Jenkin, M. R. and Harris, L. R. (2006). The subjective visual vertical and the perceptual upright, *Experimental Brain Research* **173**, 612–622.

Abstract from the 14th International Multisensory Research Forum, The Hebrew University of Jerusalem, Israel, 2013.

* Contact author: mikejcarn@gmail.com

 DOI:10.1163/22134808-000S0091

BRILL Multisensory Research 26 Supplement (2013) 126–127 brill.com/msr

Poster Presentation

Crossmodal enhancement of visual orientation discrimination by looming sounds requires functional activation of primary visual areas: A case study

Roberto Cecere [1,*], **Vincenzo Romei** [2] **and Elisabetta Ladavas** [1]

[1] University of Bologna, Italy
[2] University of Essex, UK

Abstract

Approaching or looming sounds are potentially threatening stimuli with particular impact on visual processing. The early onset of crossmodal effects by looming sounds (Romei *et al.*, 2009) and their impact on low-level visual processing (Leo *et al.*, 2011) suggests that these interactions may take place already within low-level visual cortices. To investigate the contribution of early visual areas in mediating the effects of looming sounds on visual processing, we tested a patient with bilateral occipital lesion and spared residual portions of V1/V2 (SDV). Accordingly, SDV's visual perimetry revealed blindness of the central visual field with some residual peripheral vision. We tested the influence of looming, receding and static sounds on SDV line discrimination ability in the relatively preserved visual field or in the blind visual field. We found that SDV's visual orientation sensitivity was significantly enhanced when visual stimuli were paired with looming sounds, compared to other types of sound or presented alone. Crucially, this selective crossmodal effect was found when visual stimuli were presented in the relatively preserved visual field whereas no modulation in visual orientation sensitivity was found for stimuli presented in the blind visual field. These results provide direct evidence that primary visual areas are critically involved in crossmodal modulation of visual sensitivity and that, in the presence of V1 lesion, visual pathways (e.g., retino-colliculo-extrastriate) bypassing V1 can mediate the crossmodal enhancement of basic visual abilities such as detection of spots of lights (Ladavas, 2008), but not more complex functions such as line orientation discrimination.

Keywords

Crossmodal attention, looming sounds, visual orientation

Abstract from the 14th International Multisensory Research Forum, The Hebrew University of Jerusalem, Israel, 2013.

* Contact author: roberto.cecere2@unibo.it

 DOI:10.1163/22134808-000S0092

References

Làdavas, E. (2008). Multisensory-based approach to the recovery of unisensory deficit, *Ann. N. Y. Acad. Sci.* **1124**, 98–110.

Leo, F., Romei, V., Freeman, E., Ladavas, E. and Driver, J. (2011). Looming sounds enhance orientation sensitivity for visual stimuli on the same side as such sounds, *Exp. Brain Res.* **213**(2-3), 193–201.

Romei, V., Murray, M. M., Cappe, C. and Thut, G. (2009). Preperceptual and stimulus-selective enhancement of low-level human visual cortex excitability by sounds, *Curr. Biol.* **19**, 1799–1805.

BRILL Multisensory Research 26 Supplement (2013) 128 brill.com/msr

Poster Presentation

Fast transfer of cross-modal interval training needs attentional engagement

Lihan Chen * **and Xiaolin Zhou**

Psychology Department, Peking University, China

Abstract

Here we examined whether improvement of the sub-second timing discrimination could be acquired through fast cross-modal training. To achieve this, we used a visual Ternus display (Ternus, 1926) as probe. In visual Ternus display, dependent on the inter-stimulus interval (ISI) between the two frames, participants could perceive either 'element motion' (short ISI) or 'group motion' (long ISI). In between the pre-test and post-test of discriminating the two percepts of visual Ternus (with ISI from 50 ms to 230 ms), observers received a task-relevant training, in which they were required to compare the which of the two auditory intervals or tactile intervals was longer (with auditory pitch or tactile intensity randomized). On the other hand, a second group of observers received a task-irrelevant training, in which they were required to discriminate which pair of the beeps was higher in pitch or which pair of the taps were stronger in intensity, although the time intervals in auditory/tactile pairs were tantamount to the intervals in the task-relevant situation. The results showed that after quick (about 15 mins) interval training, observers improved the sensitivity of categorizing the visual Ternus display, with a dominant percept of 'group motion', indicating subjective separation of the two frames was improved. However, this benefit was absent in the task-irrelevant condition (pitch or intensity discrimination). It suggests there might be an amodal sub-second interval representation; the transfer of interval timing training could be acquired shortly but needs task-relevant attentional engagement.

Keywords

Cross-modal training, sub-second, time interval

Abstract from the 14th International Multisensory Research Forum, The Hebrew University of Jerusalem, Israel, 2013.

* Contact author: clh20000@gmail.com

DOI:10.1163/22134808-000S0093

BRILL Multisensory Research 26 Supplement (2013) 129 brill.com/msr

Poster Presentation

Response time variability and stage dependence in the time-window-of-integration model

Hans Colonius [1,*] **and Adele Diederich** [2]

[1] Oldenburg University, Germany

[2] Jacobs University Bremen, Germany

Abstract

In the time-window-of-integration (TWIN) model framework multisensory integration occurs only if the peripheral processes elicited in different sensory channels terminate within temporal proximity of each other (Colonius and Diederich, 2004). The model has been successful in predicting mean reaction times (RTs) to crossmodal stimuli under a host of different conditions including variations of stimulus intensity, spatial and temporal correspondence, and instruction, i.e., focused attention *vs.* redundant signals paradigm (Diederich and Colonius, 2011).

Here we investigate the model's predictions about reaction time variability as measured, in particular, by RT variance. Because occurrence of multisensory interaction, that is, facilitation or inhibition, depends on the outcome of the race in the first, peripheral stage subsequent stages are not statistically independent of the duration of the first stage. Although the effect is not directly observable, it can be shown that under rather general conditions this statistical dependence is positive for multisensory inhibition and negative for facilitation. In addition, we derive explicit expressions for RT variance in the ex-Gaussian model version and also investigate effects of stimulus intensity and integration window size on RT variability.

Keywords

Time window, reaction time, variability

References

Colonius, H. and Diederich, A. (2004). *J. Cog. Neurosci.* **16**(6), 1000–1009.

Diederich, A. and Colonius, H. (2011). In: *The Neural Bases of Multisensory Processes*, M. T. Wallace and M. M. Murray (Eds), pp. 253–276. CRC Press.

Abstract from the 14th International Multisensory Research Forum, The Hebrew University of Jerusalem, Israel, 2013.

* Contact author: hans.colonius@uni-oldenburg.de

DOI:10.1163/22134808-000S0094

BRILL Multisensory Research 26 Supplement (2013) 130 brill.com/msr

Poster Presentation

A neural network model of cortical auditory–visual interactions

Cristiano Cuppini [1,*], Nadia Bolognini [2], Elisa Magosso [1], Giuseppe Vallar [2] and Mauro Ursino [1]

[1] University of Bologna, Italy
[2] University of Milano-Bicocca, Italy

Abstract
The perception of the external world is based on the integration of inputs from different sensory modalities. Recent experimental findings suggest that this phenomenon is present in lower-level cortical areas at early processing stages. The mechanisms underlying these early processes and the organization of the underlying circuitries are still a matter of debate. Here, we investigate audiovisual interactions by means of a simple neural network consisting of two layers of visual and auditory neurons. We suggest that the spatial and temporal aspects of audio-visual illusions can be explained within this simple framework, based on two main assumptions: auditory and visual neurons communicate *via* direct synapses; spatio-temporal receptive fields are different in the two modalities, auditory processing exhibiting a higher temporal resolution, visual processing a higher spatial acuity. With these assumptions, the model is able: (i) to simulate the sound-induced flash fission illusion; (ii) to reproduce psychometric curves assuming a random variability in some parameters; (iii) to account for other audio-visual illusions, such as the sound-induced flash fusion and the ventriloquism illusions. In sum, the proposed model provides a unifying summary of spatio-temporal audio-visual interactions, valuable to interpret a wide set of available findings, and a framework for future experiments.

Keywords
Neurocomputational model, multisensory integration, visual–auditory illusions

Abstract from the 14th International Multisensory Research Forum, The Hebrew University of Jerusalem, Israel, 2013.
* Contact author: cristiano.cuppini@unibo.it

 DOI:10.1163/22134808-000S0095

BRILL Multisensory Research 26 Supplement (2013) 131 brill.com/msr

MULTISENSORY RESEARCH

Poster Presentation

Distortions of the perceptual body affect basic tactile properties

Sarah D'Amour * **and Laurence R. Harris**

Department of Psychology, York University, Canada

Abstract

The internal representation of the body can be altered by artificially stimulating arm tendons which distort the body's idea of where the limbs are. If the limbs are touching another body part, then this can distort the perceived size of that body part (Lackner, 1988). Here we explore whether two-point tactile discrimination and tactile sensitivity are affected by such perceptual distortions. We used tendon vibration to distort the length of an arm and the width of the waist.

Vibrotactile stimuli were delivered with an array of 5 tactors that were spaced by 3 cms on a belt that was either laid on the arm or worn on the stomach. Two-point discrimination and tactile sensitivity detection tasks were performed using two-interval forced choice designs. The arm's perceptual length was modified by having it held by the other arm while the biceps and triceps tendons were activated by vibration. Perceived waist thickness was modified by holding the arms akimbo and stimulating the extensor or flexor tendons on both arms simultaneously (Ehrsson *et al.*, 2005).

Both two-point discrimination thresholds and sensitivity thresholds were increased when the arm was perceptually lengthened and when the waist was extended. Two-point thresholds and sensitivity thresholds were unaffected when the arm was shortened but increased when the waist was made to feel smaller.

We conclude that fundamental sensory properties determining discrimination and sensitivity thresholds are not exclusively determined by cutaneous mechanisms but depend on higher-level representations in the brain that can be distorted.

Keywords

Body representation, tactile perception, distorted body representation

References

Ehrsson, H. H., Kito, T., Sadato, N., Passingham, R. E. and Naito, E. (2005). Neural substrate of body size: Illusory feeling of shrinking the waist, *PLoS Biol.* **3**(12), e412.

Lackner, J. R. (1988). Some proprioceptive influences on the perceptual representation of body shape and orientation, *Brain* **111**, 281–297.

Abstract from the 14th International Multisensory Research Forum, The Hebrew University of Jerusalem, Israel, 2013.

* Contact author: saod16@yorku.ca

 DOI:10.1163/22134808-000S0096

BRILL Multisensory Research 26 Supplement (2013) 132 brill.com/msr

Poster Presentation

Multisensory integration, the MNS and enhanced learning

Alex Dan * **and Miriam Reiner**

The Virtual Reality and Neurocognition Lab, Dept. of Ed. in Science and Technology, Technion, Israel

Abstract

Remote interaction, especially learning and teaching became embedded in everyday functioning of corporates, educational systems and basic needs for the public. This study examines the role of multisensory integration in a touch enabled virtual world that includes physiological realistic, real-time, avatar. Interaction with the Human Realistic Avatar involves integration of auditory and visual stimuli, in the process of mimicking the motor behavior of the avatar. We ask what are the mechanisms of multisensory integration involved in the mimicking process of the avatar. In the study, we used a task with motor and cognitive components. The dependent variables were the cognitive load as measured by EEG and students' perceived cognitive load based on questionnaires. We measured the spectrum EEG using two channels.

We believe that multisensory integration of visual cues that exhibit biological motion of the avatar, result in automatic and, therefore, effortless embodiment of a mental motor plan. From a learning perspective, this might reduce the cognitive load, by leaving more working memory capacity available for processes such as elaboration or reflection on intentions of actions.

The results of the pilot study support the assumptions. Results show that there are some other EEG measures that may be associated with the activation of neuron mirror system, which should be taken in consideration in further research.

Abstract from the 14th International Multisensory Research Forum, The Hebrew University of Jerusalem, Israel, 2013.

* Contact author: alexda@technion.ac.il

DOI:10.1163/22134808-000S0097

BRILL Multisensory Research 26 Supplement (2013) 133–134 brill.com/msr

Poster Presentation

Mouth reversal extinguishes mismatch negativity induced by the McGurk illusion

Kasper Eskelund [1,2,*] **and Tobias Andersen** [1]

[1] Section for Cognitive Systems, Department of Informatics and Mathematical Modelling, Technical University of Denmark, Denmark

[2] Center for Applied Hearing Research, Department of Electrical Engineering, Technical University of Denmark, Denmark

Abstract

The sight of articulatory mouth movements (visual speech) influences auditory speech perception. This is demonstrated by the McGurk illusion in which incongruent visual speech alters the auditory phonetic percept.

In behavioral studies, reversal of the vertical mouth direction has been reported to greatly reduce the McGurk illusion (Rosenblum *et al.*, 2000). Such findings support the idea that audiovisual integration in speech to some extent relies on information regarding facial configuration.

Here we ask whether this behavioral effect is reflected in a difference in neural activity in the auditory cortex. Mismatch Negativity (MMN) is a component in the auditory Event-Related Potential (ERP) that is elicited by a change in the auditory percept. It has been shown that the McGurk illusion can induce a MMN.

We conducted an experiment in which the MMN could be elicited by the McGurk illusion induced by visual speech with either upright (unaltered) or vertically reversed mouth area.

In a preliminary analysis, we found a Mismatch Negativity component induced by the McGurk illusion for 6 of 17 participants at electrode Cz when the mouth area was upright. In comparison, these participants produced no Mismatch Negativity when the mouth was reversed. These findings mirrored behavioral findings in the same subjects of a strong McGurk response for normal audiovisual speech, which was greatly reduced for stimuli with reversed mouth area.

Keywords

Audiovisual integration, speech perception, McGurk illusion, event-related potentials, mismatch negativity

Abstract from the 14th International Multisensory Research Forum, The Hebrew University of Jerusalem, Israel, 2013.

* Contact author: kaes@dtu.dk

DOI:10.1163/22134808-000S0098

References

Näätänen, R., Gaillard, A. W. K. and Mäntysalo, S. (1978). Early selective-attention effect on evoked potential reinterpreted, *Acta Psychologica* **42**, 313–329.

Rosenblum, L. D., Yakel, D. A. and Green, K. P. (2000). Face and mouth inversion effects on visual and audiovisual speech perception, *Journal of Experimental Psychology: Human Perception and Performance* **26**, 806–819.

BRILL Multisensory Research 26 Supplement (2013) 135 brill.com/msr

Poster Presentation

Unconscious temporal association

Irune Fernández-Prieto [1,2,*]

[1] Fundació Sant Joan de Déu. Parc Sanitari Sant Joan de Déu. CIBERSAM, Spain
[2] Insitut de Recerca en Cervell, Cognició i Conducta, University of Barcelona, Spain

Abstract

To what extent does our unconscious experience with a temporal correspondence between two stimuli from different modalities influence how we subsequently perceive them? We addressed this question by exposing participants two different geometrical figures (a circle and a triangle; presented unpredictably) and a tone, for 5 min. While the circle was consistently presented 700 ms before the tone, the triangle always appeared in a completely uncorrelated fashion with respect to the tone. In two test blocks, conducted one before and the other one after the 5-min exposure phase, participants performed simultaneity judgments (SJs) regarding the circle and the exposed tone or else the triangle and the tone presented at different stimulus onset asynchronies (SOAs). The results revealed that, even if participants were unable to detect any conscious association between the stimuli presented after the 5-min exposure phase, they significantly accepted less time between the 'associated figure' (the circle) and the tone than between the 'non-associated figure' (the triangle) and the tone to judge them as being simultaneous, after exposure. Furthermore, a small but significant temporal realignment was also observed between the 2 unconsciously associated stimuli after the 5-min exposure phase. This pattern of results indicates that the preconscious establishment of perceptual links between visual and auditory stimuli modulates the subjective temporal perception of these stimuli.

Keywords

Crossmodal association, temporal adaptation, unconscious perception

Abstract from the 14th International Multisensory Research Forum, The Hebrew University of Jerusalem, Israel, 2013.

* Contact author: irunefernandez@gmail.com

 DOI:10.1163/22134808-000S0099

BRILL Multisensory Research 26 Supplement (2013) 136 brill.com/msr

Poster Presentation

Combining visual and haptic cues when judging a rod's verticality

Lindsey E. Fraser * **and Laurence R. Harris**

York University, Canada

Abstract

The Subjective Visual (SVV) and Haptic (SHV) Vertical indicate the perceived direction of gravity. Previous methods have relied on participants adjusting a rod manually, potentially introducing movement-related confounds. Here we used robust psychophysical techniques to directly compare the SVV, SHV, and a novel multimodal paradigm.

Participants judged if a probe rod (30.5 cm long, 0.9 cm diameter) was tilted to the right (clockwise) or left (counterclockwise) with respect to gravity, during whole-body tilt 30° leftwards in the roll plane. A motor set the orientation of the rod; tested orientations were chosen using a PSI adaptive staircase optimized to estimate the reliability of each judgment. For SVV, the central 5 cm of the rod was illuminated and viewed through a diffusing screen that reduced the reliability of judgments to a level comparable to SHV judgments. For SHV, participants explored the rod by touch. In a third, multimodal condition, participants used both visual and haptic cues simultaneously.

The standard deviations of the SVV, SHV and multimodal estimates were $3.9° \pm 0.2°$, $5.5° \pm 0.9°$, and $2.9° \pm 0.2°$ respectively. Across subjects, the variability and orientation of combined-cue estimates were consistent with the maximum likelihood estimate model.

We conclude that, as when comparing haptic and visual information for shape (Ernst and Banks, 2002), perceived orientation also results from optimal combination of available cues and that this estimate is available for making cross-modal judgements with the perceived direction of gravity.

Keywords

Multisensory perception, subjective vertical, perceived vertical, maximum likelihood estimate, subjective visual vertical, subjective haptic vertical, visual, haptic, sensory integration

Reference

Ernst, M. O. and Banks, M. S. (2002). Humans integrate visual and haptic information in a statistically optimal fashion, *Nature* **415**, 429–433.

Abstract from the 14th International Multisensory Research Forum, The Hebrew University of Jerusalem, Israel, 2013.

* Contact author: lfraser4@yorku.ca

 DOI:10.1163/22134808-000S0100

BRILL Multisensory Research 26 Supplement (2013) 137 brill.com/msr

Poster Presentation

Tactile feedback improves auditory spatial localization

Monica Gori [1,*], **Tiziana Vercillo** [1], **Giulio Sandini** [1] **and David Burr** [2]

[1] Robotics Brain and Cognitive Sciences Department, Istituto Italiano di Tecnologia, Genoa, Italy

[2] Dipartimento di Psicologia, Università Degli Studi di Firenze, Florence, Italy

Abstract

We recently showed that congenitally blind adults have severely impaired thresholds in an auditory spatial bisection task, pointing to the importance of vision in constructing auditory spatial maps. To explore strategies that may improve the auditory spatial sense of the blind, we investigated the impact of tactile feedback on spatial auditory localization in 29 blindfolded sighted subjects, assigned at random to one of three groups: tactile feedback ($n = 11$), verbal feedback ($n = 8$) and no feedback ($n = 10$). For all groups, auditory bisection thresholds were first measured by playing three consecutive sounds at 500 ms intervals, and subjects judged whether the second sound was spatially closer to the first or to the third. The tactile-feedback group was given two audio-tactile training sessions of 100 trials, where each auditory trial was followed by the same spatial sequence played on the subject's forearm; auditory spatial bisection thresholds were evaluated after each session. The no-feedback group did the same sequence of trials, with no feedback. In the verbal-feedback condition, the positions of the sounds was verbally reported to the subject after each training trial. Interestingly, subject performance improved significantly only after the audio-tactile training, on average by a factor of 2.5. The no-feedback sessions, and those with verbal feedback produced no statistically significant improvement. The results suggest that direct tactile feedback interacts with the auditory spatial localization system, possibly by a process of cross-sensory recalibration. More generally, the results encourage the possibility of designing rehabilitation programs to help blind individuals establish a robust sense of space.

Keywords

Tactile, audio, recalibration, spatial localization

Abstract from the 14th International Multisensory Research Forum, The Hebrew University of Jerusalem, Israel, 2013.

* Contact author: monica.gori@iit.it

 DOI:10.1163/22134808-000S0101

BRILL Multisensory Research 26 Supplement (2013) 138 brill.com/msr

Poster Presentation

Does temporal adaptation affect multisensory integration?

Vanessa Harrar [1,*], Charles Spence [1] and Laurence R. Harris [2]

[1] Department of Experimental Psychology, University of Oxford, UK
[2] Psychology Department, York University, Canada

Abstract

Generally speaking, multisensory integration is more likely to occur when the stimuli are synchronous (Stein and Meredith, 1993). Repeated exposure to temporally offset multisensory stimuli can change the perceived delay between the stimuli so that synchrony is perceived closer to the adapted delay rather than physical synchrony (Fujisaki *et al.*, 2004). If the perception of synchrony is adaptable, might the point (or delay) of maximal integration also be altered after adaptation? Temporal adaptation might be achieved by changing the processing times of the component stimuli (Harrar and Harris, 2008; Navarra *et al.*, 2009), or changing the integration mechanism. In the present study, each participant underwent daily adaptation to either synchronous or asynchronous (auditory lagging by 200 ms, or visual lagging by 60 ms) stimulus pairs. To assess unimodal processing time changes, we measured reactions times (RTs) to audio and visual stimuli after adaptation. In order to assess the effects of adaptation on multisensory integration, we measured RTs to synchronously presented AV stimuli and compared these with the RTs predicted from the Miller's race model (Miller, 1982) for each participant (Molholm *et al.*, 2004). The results comparing RTs following synchronous and asynchronous adaptation conditions are discussed in the context of perception *versus* action and current models of multisensory integration. The RTs changed considerably over a period of a week; these patterns are discussed in the context of learning to perceive synchrony.

Keywords

Reaction time, race model, adaptation, synchrony

References

Fujisaki, W., Shimojo, S., Kashino, M. and Nishida, S. (2004). *Nature Neuroscience* **7**, 773–778.
Harrar, V. and Harris, L. R. (2008). *Experimental Brain Research* **186**, 517–524.
Miller, J. O. (1982). *Cognitive Psychology* **14**, 247–279.
Molholm, S., Ritter, W., Javitt, D. C. and Foxe, J. J. (2004). *Cerebral Cortex* **14**, 52–465.
Navarra, J., Hartcher-O'Brien, J., Piazza, E. and Spence, C. (2009). *PNAS* **106**, 9169–9173.
Stein, B. E. and Meredith, M. A. (1993). *Merging of the Senses*. MIT Press, Cambridge, MA.

Abstract from the 14th International Multisensory Research Forum, The Hebrew University of Jerusalem, Israel, 2013.

* Contact author: vanessa.harrar@psy.ox.ac.uk

DOI:10.1163/22134808-000S0102

BRILL Multisensory Research 26 Supplement (2013) 139 brill.com/msr

Withdrawn Presentation

A sense of time: Non-linear distortions in perceived time across the senses

Jess Hartcher-O'Brien [1,*], **Massimilliano Di Luca** [2] **and Marc O. Ernst** [3]

[1] LIMSI CNRS, France
[2] University of Birmingham, UK
[3] University of Bielefeld, CITEC, Germany

Abstract

Perceived time is not veridical but distorted and differs across the senses. Here we ask, which points in the perception of a temporal event contribute most significantly to these multisensory distortions? To this end, we investigated perceptual estimates of temporal landmarks (onset, peak amplitude, and offset) for a Gaussian standard signal. Furthermore, we identified that the perception of these landmarks differ across vision and audition. Participants were asked to compare the onset, peak and offset landmarks of the standard long visual/auditory stimuli to short spike-like stimuli in vision or audition. Four combinations were tested: V–v, A–a, V–a, A–v. Results demonstrate that the visual as compared to auditory standard stimuli were perceived as shorter. More interestingly, we found a compression in the perceived duration for onset-peak intervals compared to peak-offset intervals. This compression effect was more pronounced in the visual modality. We compared the distortion results to those obtained with a second signal envelope type and demonstrated the same pattern of distortions for this temporally more veridical. However, the delayed onset, advanced peak and offset were the same across the two signals. The differences in perceived duration can potentially be used to explain multisensory illusions such as the flash lag effect and perceived crossmodal asynchronies. Such distortions can be accounted for by simulated neural response functions.

Keywords

Temporal discrepancies, auditory visual plasticity, non-veridical perception

---

Abstract from the 14th International Multisensory Research Forum, The Hebrew University of Jerusalem, Israel, 2013.

* Contact author: jhartcher@tuebingen.mpg.de

 DOI:10.1163/22134808-000S0103

BRILL Multisensory Research 26 Supplement (2013) 140–141 brill.com/msr

MULTISENSORY RESEARCH

Poster Presentation

Visual allocentric reference frames effects on audio-visual spatial fusion

David Hartnagel *, **Alain Bichot, Patrick Sandor and Corinne Roumes**

ACSO–IRBA, France

Abstract

Variation over space of Audio-Visual (AV) spatial fusion has been investigated in darkness and light conditions (Hartnagel *et al.*, 2007; Roumes *et al.*, 2004). Those experiments revealed a gaze shift effect, indicating a reference frame of AV fusion space being neither head- nor eye-centered. Results in vision research have shown influence of visual allocentric reference frame on visual localization. Schmidt *et al.* (2003) have shown local distortion effect of visual landmark and experiments on the Reolof effect have shown shift of localization relative to the asymmetric surrounding display (Dassonville *et al.*, 2004). Our experiment investigates visual allocentric effects on AV fusion. An hemi-cylindrical screen hiding 21 loudspeakers in a 2D arrangement, a projector displays on the screen a green permanent rectangular large background (135°H × 80°V); participant (head and body aligned) was sideways-oriented so the display appeared shifted 15° to the right relative to straight ahead and the surrounding visual frame was asymmetrical (frame offset). In each trial a vertical line providing a visual landmark randomly either straight ahead (head 0°) or 15° to the right (mid-display), a broadband noise burst and a 1° spot of light, 500 ms duration, were simultaneously presented with random 2D spatial disparity. The task was about perception of spatial unity of the bimodal stimulus (fusion). Results showed that AV fusion depends mainly on egocentric reference frames relative position (gaze and head) and that local allocentric reference frame has no significant effect. Comparisons with previous results confirm the importance of surrounding visual display.

Keywords

Audio, visual, fusion, reference frames, space

Abstract from the 14th International Multisensory Research Forum, The Hebrew University of Jerusalem, Israel, 2013.

* Contact author: davir.hartnagel@irba.fr

 DOI:10.1163/22134808-000S0104

References

Dassonville, P., Bridgeman, B., Kaur Bala, J., Thiem, P. and Sampanes, A. (2004). The induced Roelofs effect: two visual systems or the shift of a single reference frame? *Vision Research* **44**(6), 603–611.

Hartnagel, D., Bichot, A. and Roumes, C. (2007). Eye position affects audio-visual fusion in darkness, *Perception* **36**(10), 1487–1496.

Roumes, C., Hartnagel, D. and Godfroy, M. (2004). Visual–auditory fusion and frames of reference, *Perception* **33**.

Schmidt, T., Werner, S. and Diedrichsen, J. (2003). Spatial distortions induced by multiple visual landmarks: How local distortions combine to produce complex distortion patterns, *Perception & Psychophysics* **65**(6), 861.

BRILL Multisensory Research 26 Supplement (2013) 142 brill.com/msr

Poster Presentation

The impact of saliency on overt visual selection in early-deaf adults

Benedetta Heimler [1,*], **Wieske van Zoest** [1], **Francesca Baruffaldi** [1], **Pasquale Rinaldi** [2], **Maria Cristina Caselli** [2] **and Francesco Pavani** [1]

[1] University of Trento, Italy
[2] National Research Council, Italy

Abstract

Following bilateral deafness, vision reorganizes to overcome the lack of audition. Results from covert attention studies show enhanced reactivity and rapid attention orienting to abrupt visual stimuli in deaf people, suggesting that crossmodal changes may prioritize saliency-driven attention capture in this population. However, deaf individuals also need to maintain goal-directed behaviors. We investigated the time-course of overt visual selection in deaf adults, to assess when in time deaf individuals implement efficient top-down control over saliency-driven search. Eighteen early-deaf adults and sixteen hearing controls performed an oculomotor additional singleton paradigm. For deaf participants language abilities in oral and sign languages were assessed. Participants made a speeded saccadic eye movement to a unique orientation singleton. The target was presented among homogenous non-targets and one additional orientation singleton that was more, equally or less salient than the target. Results showed a similar pattern of performance in the two groups: fast initiated saccades were saliency-driven whereas later initiated saccades were more goal-driven. However, deaf were overall slower than hearing participants at initiating saccades. This delay in saccadic latencies was driven by a sub-group of deaf adults, who were able to bypass saliency capture by withholding their saccades until target selection was efficiently achieved. No correlation with linguistic abilities emerged, showing no role of sign language acquisition in this behavior. Taken together, the present findings reveal that overt visual reorganization in deaf adults does not drastically alter the interplay between bottom-up and top-down strategies. If anything, it prioritizes the implementation of efficient goal-directed behaviors.

Keywords

Deafness, eye-movements, visual selection

Abstract from the 14th International Multisensory Research Forum, The Hebrew University of Jerusalem, Israel, 2013.

* Contact author: benedetta.heimler@unitn.it

 DOI:10.1163/22134808-000S0105

BRILL Multisensory Research 26 Supplement (2013) 143 brill.com/msr

Poster Presentation

Stable and dynamic, context dependent elements of sensory processing

Uri Hertz * **and Amir Amedi**

The Hebrew University of Jerusalem, Israel

Abstract

Classic view of multisensory perception relies on independent processing of sensory inputs in sensory specific cortices, and integration of multiple inputs in a bottom-up, deterministic manner. This rigid, hierarchical structure is being challenged by evidence of multisensory responses in areas which were thought to be sensory specific. However, these evidence show high variability, with some studies reporting crossmodal enhancements, some observe crossmodal inhibition, and others fail to observe any crossmodal effects. Here we seek to examine whether the source of this variability is the dynamic, context dependent nature of multisensory perception. Subjects were scanned during three audiovisual experiments which differ in task, experience and the information conveyed by the auditory input while keeping the stimuli intact. This was enabled by teaching the subjects a novel visual to auditory sensory substitution algorithm (SSA), originally developed as a rehabilitation tool for the blind. Our results reveal two elements of multisensory perception — a stable, deterministic element and a fast, dynamic context dependent element. Sensory cortices did not change their sensory preference throughout the experiments. However, crossmodal attenuations of sensory specific areas were also detected. These changed their direction completely after only a short period of learning SSA, from visual attenuations of primary auditory cortex to auditory attenuations of visual cortex. Outside sensory areas a number of cortical regions, mostly left lateralized, changed their sensory preference, shifting it from one sensory modality to the other, while others demonstrated constant multisensory preference. We therefore suggest that in multisensory perception there are both stable and dynamic processes, allowing stability in one hand and a fast, dynamic tuning of the system when required, and both are taking place together, which may lead to seemingly contradicting evidence. In order to tweeze them apart, and overcome these contradictions, one has to take into account not only on the stimuli themselves, but rather on the context in which they are perceived, and the information they convey.

Keywords

fMRI, multisensory perception, audiovisual

Abstract from the 14th International Multisensory Research Forum, The Hebrew University of Jerusalem, Israel, 2013.

* Contact author: uri.hertz@mail.huji.ac.il

 DOI:10.1163/22134808-000S0106

BRILL Multisensory Research 26 Supplement (2013) 144 brill.com/msr

Poster Presentation

Access to art: Different ways of looking

Abigail Hirsch *

Royal Academy of Arts, Israel

Abstract

The goal is wherever possible to remove or reduce physical, sensual, and Intellectual barriers to access and to ensure that all aspects of art galleries, exhibitions and activities are accessible as possible for adults and children with mild disabilities to profound and multiple learning disabilities (PMLD).

In this presentation I will review the experience of the children while visiting at an exhibition and taking part of a creative session, being exposed to multi sensory (everyday life) props and using facilities, which are being adapted to their abilities.

I will discuss the engagement of the children through sensory activities and/or stimulations placing with in the art context followed by questions raised in regarding to their perceptions and experiences of the artwork and/or activity. For children with PMLD taking part in an arts-based activity, a visit to the gallery goes beyond the art work and include the impact of new physical surroundings; echoes in different rooms is very exciting for some children and can prompt exceptional responses. New experiences, such as looking at a striking artwork, listening to a soundscape or making work, invite these children to respond directly to what they are seeing, hearing, smelling or doing.

The experience facilitates new ways to communicate and can help the children to better understand the world around them. Fundamental to using creative arts to communicate with children with PMLD is the opportunity to develop a set of arts-related experiences that incorporate a range of senses; visual stimulations, temperature, touch and smell are all powerful ways to reach children who face a complex set of communication challenges.

Abstract from the 14th International Multisensory Research Forum, The Hebrew University of Jerusalem, Israel, 2013.

* Contact author: abihirsch@gmail.com

DOI:10.1163/22134808-000S0107

BRILL Multisensory Research 26 Supplement (2013) 145 brill.com/msr

Poster Presentation

Temporal expectation and spectral expectation operate in distinct fashion on neuronal populations

Yi-Fang Hsu [1,*], Jarmo A. Hämäläinen [2] and Florian Waszak [1]

[1] Laboratoire Psychologie de la Perception (UMR 8158), Université Paris Descartes & CNRS, France

[2] Department of Psychology, University of Jyväskylä, Finland

Abstract

The formation of temporal expectation (i.e., the prediction of 'when') is of prime importance to sensory processing. It can modulate sensory processing at early processing stages probably *via* the entrainment of low-frequency neuronal oscillations in the brain. However, sensory predictions involve not only temporal expectation but also spectral expectation (i.e., the prediction of 'what'). Here we investigated how temporal expectation may interrelate with spectral expectation by explicitly setting up temporal expectation and spectral expectation in a target detection task. We found that temporal expectation and spectral expectation interacted on reaction time (RT). RT was shorter when targets were temporally expected than when they were temporally unexpected. Moreover, such temporal expectation effect was larger with than without spectral expectation. However, this interaction in the behavioural data did not result from an interaction in the electroencephalography (EEG), where we observed independent main effects of temporal expectation and spectral expectation. More precisely, we found that early event-related potential (ERP) components and the entrainment of low-frequency neuronal oscillations were exclusively modulated by temporal expectation, whereas only the late ERP component was modulated by spectral expectation. Our results, thus, support the idea that temporal expectation and spectral expectation operate in distinct fashion on neuronal populations.

Abstract from the 14th International Multisensory Research Forum, The Hebrew University of Jerusalem, Israel, 2013.

* Contact author: yi-fang.hsu@parisdescartes.fr

DOI:10.1163/22134808-000S0108

BRILL Multisensory Research 26 Supplement (2013) 146–147 brill.com/msr

MULTISENSORY RESEARCH

Poster Presentation

Expectation of tactile events is enough to induce crossmodal congruency effect (CCE)

Federica Iezzi [1,*], **Pasquale Cardellicchio** [1], **Ettore Ambrosini** [1], **Francesca Ferri** [2] **and Marcello Costantini** [1]

[1] University G. d'Annunzio, Chieti, Italy

[2] Mind, Brain Imaging and Neuroethics, Royal Ottawa Healthcare, Institute of Mental Health Research, Ottawa, Canada

Abstract

New approaches understand sensory processing as constructive and active in nature. Expectations about upcoming sensory events can be utilized to prepare sensory cortices by instantiating a neural context that allows for enhanced processing of the forthcoming event. But, how deeply is expectation involved in sensory processing? One of the most useful ability of our brain is to integrate input from different sensory modalities in order to create a coherent representation of the environment. Does expectation have a role to play in basic crossmodal integration? In this study we tackled this issue by taking advantage of a Crossmodal Congruency Task.

Participants were stimulated at the index or thumb of the right hand, using solenoids mounted on a foam cube. Simultaneously they received a visual stimulus adjacent to the active solenoid (congruent) or adjacent to the inactive solenoid (incongruent). The task was to respond to the elevation of the visual stimulus while ignoring the tactile distractor. To induce expectation over the tactile stimulus, in 80% of the trials either a high or a low pitched tone preceded the stimulation of the index and thumb finger, respectively. In the remaining 20% of the trials the tactile stimulus was not delivered.

Results fully replicated basic CCE. Strikingly, CCE was observed, though at a less degree, also when the tactile stimulus was not delivered but merely expected.

Our findings suggest that a coherent representation of the environment, as achieved by crossmodal integration, does not require real stimuli. Expected stimuli are enough to this aim.

Keywords

Crossmodal congruency task, predictive coding, tactile expectation

---

Abstract from the 14th International Multisensory Research Forum, The Hebrew University of Jerusalem, Israel, 2013.

* Contact author: federica_iezzi@hotmail.it

 DOI:10.1163/22134808-000S0109

References

Engel, A. K., Fries, P. and Singer, W. (2001). Dynamic predictions: Oscillations and synchrony in top-down processing, *Nature Reviews Neuroscience* **2**, 704–716.

Friston, K. (2010). Is the free-energy principle neurocentric? *Nat. Rev. Neurosci.* **11**, 605.

Spence, C., Pavani, F., Maravita, A. and Holmes, N. (2004). Multisensory contributions to the 3-D representation of visuotactile peripersonal space in humans: evidence from the crossmodal congruency task, *J. Physiol. Paris* **98**, 171–189.

BRILL Multisensory Research 26 Supplement (2013) 148 brill.com/msr

Poster Presentation

A body illusion by movements or by touch: Comparing different ways of inducing the rubber hand illusion

Andreas Kalckert * **and Henrik Ehrsson**

Brain, Body and Self Laboratory, Dep. of Neuroscience, Karolinska Institutet, Sweden

Abstract

The rubber hand illusion is a bodily illusion where a model hand is experienced as part of the own body. The classical version of this illusion is based on synchronous visuo-tactile stimulation applied to the model in view and the person's real real hand, which is hidden from view. However it has been also shown that the illusion can be induced by passive or active finger movements, without any tactile stimulation by an external object.

In the present study participants experience active movements, passive movements and visuo-tactile stimulation of a model hand. Although these situations represent different combinations of sensory information, they all create a sense of ownership towards the model hand. This raises the question which of these combinations is most potent in eliciting the illusion. In particular one may ask if the combination of afferent and efferent signals in active movements would result in the same experience as passive visuo-tactile stimulation.

We present data from subjective and objective measures of the illusion and show here that the illusion seems to be similar, regardless if it is induced by synchronized visuo-tactile stroking, active finger movements or passive finger movements. These observations show that different combinations of sensory input can all lead to a very similar phenomenological experience and exemplify again that the illusion can be induced by any combination of multisensory information, and not by a specific combination alone.

Keywords

Multisensory integration, sensorimotor integration, body perception, self-recognition

Abstract from the 14th International Multisensory Research Forum, The Hebrew University of Jerusalem, Israel, 2013.

* Contact author: andreas.kalckert@ki.se

 DOI:10.1163/22134808-000S0110

BRILL Multisensory Research 26 Supplement (2013) 149–150 brill.com/msr

Poster Presentation

Rotating straight ahead or translating in circles: How we learn to integrate contradictory multisensory self-motion cue pairings

Mariia Kaliuzhna [1,*], **Mario Prsa** [2] **and Olaf Blanke** [1,3]

[1] Laboratory of Cognitive Neuroscience, Brain Mind Institute, Ecole Polytechnique Fédérale de Lausanne, Lausanne, Switzerland

[2] Université de Genève, Neural Circuits and Behavior, Switzerland

[3] Center for Neuroprosthetics, Ecole Polytechnique Fédérale de Lausanne, Lausanne, Switzerland

Abstract

Humans integrate multisensory information to reduce perceptual uncertainty when perceiving the world (Hillis *et al.*, 2002, 2004) and self (Butler *et al.*, 2010; Prsa *et al.*, 2012) and it has been shown that two multisensory cues are combined and give rise to a single percept only if attributed to the same causal event (Koerding *et al.*, 2007; Parise *et al.*, 2012; Shams and Beierholm, 2010). A growing body of literature studies the limits of such integration for bodily self-consciousness and the perception of self-location under normal and pathological conditions (Ionta *et al.*, 2011). We extend this research by investigating whether human subjects can learn to integrate two arbitrary visual and vestibular cues of self-motion due to their temporal co-occurrence.

We conducted two experiments ($N = 8$ each) in which whole-body rotations were used as the vestibular stimulus and optic flow as the visual stimulus. The vestibular stimulus provided a yaw self-rotation cue, the visual — a roll (experiment 1) or pitch (experiment 2) rotation cue. Subjects made a relative size comparison between a standard rotation size and a variable test rotation size. Their discrimination performance was fit with a psychometric function and perceptual discrimination thresholds were extracted. We compared experimentally measured thresholds in the bimodal condition with theoretical predictions derived from the single cue thresholds.

Our results show that human subjects can learn to combine and optimally integrate vestibular and visual information, each signaling self-motion around a different rotation axis (yaw *versus* roll as well as pitch). This finding suggests that the experience of two temporally co-occurring but spatially unrelated self-motion cues leads to inferring a common cause to these two initially unrelated sources of information about self-motion.

Abstract from the 14th International Multisensory Research Forum, The Hebrew University of Jerusalem, Israel, 2013.

* Contact author: mariia.kaliuzhna@epfl.ch

 DOI:10.1163/22134808-00S0111

Keywords
Multisensory integration, self-motion, visuo-vestibular

References

Butler *et al.* (2010).
Hillis *et al.* (2002).
Hillis *et al.* (2004).
Ionta *et al.* (2011).
Koerding *et al.* (2007).
Parise *et al.* (2012).
Prsa *et al.* (2012).
Shams and Beierholm (2010).

BRILL Multisensory Research 26 Supplement (2013) 151 brill.com/msr

Poster Presentation

Cortical dynamics during rubber hand illusion

Noriaki Kanayama [1,*], Alberto Morandi [2], Daniele Didino [2], Stefania Cannella [2], Kazuo Hiraki [3] and Francesco Pavani [2]

[1] École Polytechnique Fédáerale de Lausanne, Switzerland
[2] University of Trento, Italy
[3] The University of Tokyo, Japan

Abstract
Rubber hand illusion is an epoch-making phenomenon on the study about the body ownership. The neural correlates of this illusory phenomenon has been revealed by many researchers, while the information flow in the cortex related this illusion was not clear.

We recorded EEG during simultaneous presentation of visuotactile stimuli using a rubber hand with a setting inducing rubber hand illusion, and also obtain the subjective reports for illusory feeling. Using renormalized partial directed coherence (RPDC) as an index of a cortical causality, we evaluated the information flow inducing the illusory experiences. As a result, at the low frequency band, we found the significant increase of neural connectivity between the parietal area and the other cortical areas. By correlation analysis, we found that some causalities are correlated with the rubber hand illusion and proprioceptive drift.

Keywords
EEG, causality analysis, rubber hand illusion

Abstract from the 14th International Multisensory Research Forum, The Hebrew University of Jerusalem, Israel, 2013.
* Contact author: noriaki.kanayama@epfl.ch

 DOI:10.1163/22134808-000S0112

BRILL Multisensory Research 26 Supplement (2013) 152 brill.com/msr

Poster Presentation

Accurate and precise parameter recovery with the basic twin model

Farid I. Kandil *

Jacobs University Bremen, SHSS Psychology, Germany

Abstract

The 'time window of integration' (TWIN) model (e.g., Colonius and Diederich, 2004; Diederich and Colonius) allows to predict response speed effects in multisensory settings. E.g., in the focused attention paradigm (FAP), subjects are instructed to respond to stimuli of the target modality only, yet reaction times are shorter if the unattended stimulus is presented in a certain temporal relation to the target stimulus. The TWIN model accounts for these cross-modal effects by proposing that all the initially unimodal bits of information must arrive at the point of integration within a certain time window in order to be integrated and thus to allow response enhancements like the observed reaction time reductions. It has been successfully used to account for empirical data. Here, we conduct a parameter recovery study of the basic TWIN model with five parameters for the duration of the visual and acoustic unimodal and the integrated second stage, the length of the time window, and the size of the effect. We conducted 576 'experiments' with different parameter value sets, each comprising 500 'subjects', for which data for 80 trials in a FAP setting were generated, averaged and fed into the recovery process. Parameter estimations were evaluated in terms of absolute and relative accuracy and precision. Results show that deviations from the true value are of only insignificant size for all parameters. Especially duration parameters for the unimodal stage of the focused stimulus and the integrated second stage are both highly accurate and precise, in fact to such an extent that they match statistics of single cell recordings.

Keywords

Model, reaction times, focuse attention paradigm (FAP), time window, parameter recovery study

References

Colonius, H. and Diederich, A. (2004). Multisensory interaction in saccadic reaction time: A time-window-of-integration model, *J. Cogn. Neurosci.* **16**, 1000–1009.

Diederich, A. and Colonius, H. (2008). Crossmodal interaction in saccadic reaction time: Separating multisensory from warning effects in the time window of integration model, *Exp. Brain Res.* **186**, 1–22.

Abstract from the 14th International Multisensory Research Forum, The Hebrew University of Jerusalem, Israel, 2013.

* Contact author: f.kandil@jacobs-university.de

 DOI:10.1163/22134808-000S0113

BRILL *Multisensory Research 26 Supplement (2013) 153* brill.com/msr

Poster Presentation

Motor-movements driving visual motion perception

Mirjam Keetels * **and Jean Vroomen**

Tilburg University, The Netherlands

Abstract

Recent studies have shown that auditory signals can induce motion perception of a static visual stimulus, or 'Sound-Induced-Visual-Motion' [SIVM; Hidaka *et al.*, 2009, 2011]. Here we explore in two experiments whether motor movements can also change perceived visual motion. In Experiment 1, participants judged the direction of two flashing horizontal bars that either moved upward or downward (with 0.1; 0.4; 0.7 or 1.0 deg vertical bar displacement). For testing the effect of co-occurring motor movements of the hand, participants were instructed to make two button-presses, either upward (right thumb on lower button followed by right index on upper button) or downward (up-down), and the visual bars were presented at the timing of the button-presses. The results demonstrate overall more upward responses when the motor movement was upward and more down-ward responses when moving downward confirming a 'Motor-Induced-Visual-Motion' (MIVM). In Experiment 2, the role of response bias was excluded in a d′prime paradigm in which each trial consisted of 6 flashing vertical bars that were either moving (left–right–left, etc.) or static. Participants judged whether the bar was moving or static. Participants performed a moving motor-movement (left hand index button press followed by right hand index button press), a static motor movement (pressing twice with left or right index) or no motor movement (visual condition). Results showed more illusory motion (lower d′prime) when participants performed a moving motor movement as compared to the static or no motor movement conditions. Together these two experiments demonstrate Motor-Induced-Visual-Motion, showing that motor movements can affect ambiguous visual motion perception.

Keywords

Motor-movements, visual motion perception

References

Hidaka, S., *et al.* (2009). Alternation of sound location induces visual motion perception of a static object, *PLoS One* **4**(12).

Hidaka, S., *et al.* (2011). Auditory motion information drives visual motion perception, *PLoS One* **6**(3).

Abstract from the 14th International Multisensory Research Forum, The Hebrew University of Jerusalem, Israel, 2013.

* Contact author: m.n.keetels@uvt.nl

 DOI:10.1163/22134808-000S0114

BRILL Multisensory Research 26 Supplement (2013) 154–155 brill.com/msr

Poster Presentation

Audio-visual interactions in the perception of intention from actions

Hanni Kiiski [1,*], **Ludovic Hoyet** [2], **Katja Zibrek** [2], **Carol O'Sullivan** [2] **and Fiona N. Newell** [1]

[1] School of Psychology and Institute of Neuroscience, Trinity College Dublin, Ireland
[2] Graphics, Vision and Visualisation Group, School of Computer Science and Statistics, Trinity College Dublin, Ireland

Abstract

Although humans can infer other people's intentions from their visual actions (Blakemore and Decety, 2001), it is not well understood how auditory information can influence this process. We investigated whether auditory emotional information can influence the perceived intention of another from their visual body motion. Participants viewed a set of videos which presented point light displays (PLDs) of 10 actors (5 male) who were asked to portray a 'hero' or a 'villain' character. Based on a 2-AFC design, participants categorised each visual character as having 'good' or 'bad' intentions. Response accuracy and speed were recorded. Performance on visual-only trials exceeded chance performance suggesting that participants were efficient at judging intentions from PLDs. We then paired auditory vocal stimuli which were associated with either positive (happy) or negative (angry) emotions with each of the PLDs. The auditory stimuli were taken from Belin *et al.* (2008) and consisted of non-verbal bursts ('ah') recorded from 10 actors (5 male). Each vocalisation was randomly paired with a sex-matched PLD (60 PLD-voice combinations). We found that both the categorisation responses and the speed of those responses were affected by the inclusion of the auditory stimuli. Specifically, reaction times were facilitated when the auditory emotion (positive or negative) matched the perceived intentions (good or bad respectively) relative to unisensory conditions. Our findings suggest important interactions between audition and visual actions in perceiving intentions in others and are consistent with previous findings of audio-visual interactions in action-specific visual regions of the brain (e.g., Barraclough *et al.*, 2005).

Keywords

Audio-visual interaction, intention perception, action perception

---

Abstract from the 14th International Multisensory Research Forum, The Hebrew University of Jerusalem, Israel, 2013.

[*] Contact author: hkiiski@tcd.ie

DOI:10.1163/22134808-000S0115

References

Barraclough, N. E., Xiao, D., Baker, C. I., Oram, M. W. and Perrett, D. I. (2005). Integration of visual and auditory information by superior temporal sulcus neurons responsive to the sight of actions, *J. Cogn. Neurosci.* **17**(3), 377–391.

Belin, P., Fillion-Bilodeau, S. and Gosselin, F. (2008). The Montreal Affective Voices: A validated set of nonverbal affect bursts for research on auditory affective processing, *Behavior Research Methods* **40**(2), 531–539.

Blakemore, S.-J. and Decety, J. (2001). From the perception of action to the understanding of intention, *Nature Reviews* **2**, 561–567.

BRILL Multisensory Research 26 Supplement (2013) 156 brill.com/msr

Poster Presentation

Multisensory ambiguities in impossible worlds: Assessing auditory, visual, and motor contributions to the representation of space

Thorsten Kluss *, **Niclas Schult, Tim Hantel, William E. Marsh and Christoph Zetzsche**

Cognitive Neuroinformatics, Bremen University, Germany

Abstract

We conducted behavioral experiments on visual, auditory, and motor contributions to the human representation of space in virtual reality environments using an 'impossible-worlds paradigm'. The experiments were run with an omnidirectional locomotion input device, the 'Virtusphere', which is a rotatable 10-foot hollow sphere that allows a subject inside to walk in any direction for any distance, while immersed in a virtual environment. Both the rotation of the sphere and the movement of a subject's head were tracked to process the subject's view within the virtual environment presented on a head-mounted display. Auditory features were dynamically processed in order to exactly align sound sources and visual objects. Using this experimental setup the subjects were presented with 'impossible worlds', i.e., virtual environments with geometrical and topological properties, which are physically not possible. In previous experiments we have shown that subjects are able to navigate inside these impossible worlds (Zetzsche *et al.*, 2009), despite the fact that different interpretations of their spatial structure are in conflict, since there is no single (physically plausible) interpretation accounting for all sensory perceptions of the subjects.

In the present study we manipulated these physically 'impossible' properties either in the visual or in the auditory domain (so that each modality supports one of the possible interpretations) and assessed how these manipulations affected the subjects' internal representations of space. We discuss our results with respect to auditory, visual, and motor contributions to the internal spatial representation, the interaction of modalities, and the implication on the notion of motor action as a linking element between the senses.

Keywords

Audiovisual, spatial cognition, sensorimotor

Reference

Zetzsche, C., Galbraith, C., Wolter, J. and Schill, K. (2009). Representation of space: image-like or sensorimotor? *Spatial Vision* **22**(5), 409–424.

Abstract from the 14th International Multisensory Research Forum, The Hebrew University of Jerusalem, Israel, 2013.

* Contact author: tox@uni-bremen.de

 DOI:10.1163/22134808-000S0116

BRILL Multisensory Research 26 Supplement (2013) 157–158 brill.com/msr

Poster Presentation

Amplified extrastriate visual cortical projections to the dorsal zone of auditory cortex following early- and late-onset deafness

Melanie A. Kok *, **Nicole Chabot and Stephen G. Lomber**
University of Western Ontario, Canada

Abstract
Adaptive cross-modal plasticity refers to cortical reorganization that takes place across sensory modalities following the loss of a sensory system. This loss of one modality is often accompanied by a heightening of the remaining senses (Bavelier and Neville, 2002). Cat auditory cortex has been shown to undergo cross-modal reorganization following deafness, such that deaf animals are better able to detect visual motion than hearing controls, an advantage which can be localized to the dorsal zone (DZ) of auditory cortex (Lomber *et al.*, 2010). However, because the structural adaptations that might subserve this plasticity remain largely unknown, the purpose of the present investigation was to examine modifications in anatomical connectivity following early- or late-onset deafness. We injected DZ of hearing, early- and late-deafened cats ($n = 5$ per group) with biotinylated dextran amine (BDA). Immunohistochemistry was performed following perfusion to reveal the tracer. Overall, the pattern of cortical projections to DZ was similar in both early- and late-deafened cats and hearing controls. However, an increase in projection strength from visual areas involved in visual motion processing was observed in both early- and late-deafened animals compared to hearing controls. Our observations suggest that amplified cortical projections from visual motion processing regions to DZ may underlie the cross-modal reorganization that functionally manifests as superior visual motion detection abilities in the deaf animal. Furthermore, the degree to which this reorganization occurs is similar between early- and late-deafened animals, suggesting that structural compensations can occur regardless of duration of acoustic experience.

Keyword
Neuroanatomy

Abstract from the 14th International Multisensory Research Forum, The Hebrew University of Jerusalem, Israel, 2013.
* Contact author: mkok@uwo.ca

 DOI:10.1163/22134808-000S0117

References

Bavalier, D. and Neville, H. J. (2002). Cross-modal plasticity: Where and how? *Nature Reviews Neuroscience* **3**, 443–452.

Lomber, S. G., Meredith, M. A. and Kral, A. (2010). Cross-modal plasticity in specific auditory cortices underlies visual compensations in the deaf, *Nature Neuroscience* **13**, 1421–1427.

BRILL Multisensory Research 26 Supplement (2013) 159 brill.com/msr

Poster Presentation

When the brain fails to recalibrate audiovisual simultaneity: Hysteresis in synchrony perception

Anne Kösem [1,*], **Jean-Rémy Martin** [2] **and Virginie van Wassenhove** [1]

[1] INSERM-CEA Cognitive Neuroimaging Unit, France

[2] Institut Jean Nicod, IEC-ENS, France

Abstract

To accurately perceive the synchrony of visual and auditory events, the brain has to compensate for the speed differences between light and sound. This compensation may be accomplished by the mechanism of temporal recalibration (Fujisaki *et al.*, 2004; Vroomen *et al.*, 2004) a mechanism through which the subjects become more tolerant to constant audiovisual (AV) asynchrony by virtue of adaptation. However, a constant AV lag means that both the source and the observer are fixed in space; in an ecological setting, if a source is moving backward or away from the observer it will induce a progressive change in the AV delay. The present study shows that synchrony perception of slowly synchronizing AV stimuli is driven by a persistence effect or hysteresis (Hock *et al.*, 1997). Specifically, when AV stimuli are progressively synchronizing, participants do not compensate for the initial asynchrony but rather persist in perceiving asynchrony; surprisingly however, slowly desynchronizing stimuli do not alter the perception of synchrony. Our results show that synchrony perception strongly depends on the dynamics of past AV stimulation. Whereas temporal recalibration effects suggest that the brain has a natural tendency to compensate for AV time lags emitted from a distant stationary source, our results suggest that compensation of AV lags for a moving source may not occur and perceived asynchrony persists.

Keywords

Simultaneity, audiovisual, hysteresis, multisensory, adaptation, recalibration, psychophysics

References

Fujisaki, W., Shimojo, S., Kashino, M. and Nishida, S. (2004). Recalibration of audiovisual simultaneity, *Nature Neuroscience*.

Hock, H. S., Kogan, K. and Espinoza, J. K. (1997). Dynamic, state-dependent thresholds for the perception of single-element apparent motion: Bistability from local cooperativity, *Perception & Psychophysics*.

Vroomen, J., Keetels, M., de Gelder, B. and Bertelson, P. (2004). Recalibration of temporal order perception by exposure to audio-visual asynchrony, *Cognitive Brain Research*.

Abstract from the 14th International Multisensory Research Forum, The Hebrew University of Jerusalem, Israel, 2013.

* Contact author: kosem.anne@gmail.com

 DOI:10.1163/22134808-000S0118

BRILL Multisensory Research 26 Supplement (2013) 160–161 brill.com/msr

Poster Presentation

Unleashing synaesthetic-like effects by stimulating the DLPFC but not PPC

Flor Kusnir *
University of Glasgow, UK

Abstract
Models on the mechanisms underlying letter–colour synaesthesia diverge on a central question: whether triggered sensations reflect (a) quantitatively or (b) qualitatively deviant brain organization. We previously found evidence for (a) after observing synaesthesia-like letter–colour binding in adult non-synaesthetes following execution of a visual letter search task which employed likelihood manipulations of letter–colour pairings to implicitly train letter–colour associations (Kusnir and Thut, 2012). The newly-formed associations were synaesthesia-like, since correlating with the synaesthetic-Stroop and showing the colour-opponency effect, present in synaesthetes (Nikolić *et al.*, 2007). This latter effect manifests as increased synaesthetic-Stroop interference when the real colour of a letter is opponent to the synaesthetic/associated colour, in line with involvement of visual/colour areas in letter–colour binding.

Here, we investigated the brain areas involved in the formation of these synesthetic associations. Based on (Cohen Kadosh and Iuculano, 2011) revealing reciprocal involvement of the PPC and dlPFC in distinct aspects of learning in a numerical conception task, we hypothesized that these two areas may also be differentially implicated in the learning of synaesthesia-like letter–colour associations by non-synaesthetes. Previous results would predict that while PPC supports processing of these associations after learning (see also Esterman *et al.*, 2006), dlPFC shows a reciprocal function, i.e., suppressing these associations under normal conditions.

Using bilateral tDCS, we interfered with either PPC or dlPFC in two groups while they performed the letter–colour association task as in (Kusnir and Thut, 2012). A third group performed the same task but without tDCS (control group). All three groups used non-opponent colour pairs for letter–colour association learning, to avoid ceiling effects. In comparison to the control group, dlPFC-stimulation significantly enhanced letter–colour binding. This enhancement was substantial, leading to interference between learned and real colours in the order of the colour-opponency effect (despite the use of non-opponent colour pairs). No such effect was observed with PPC stimulation.

This provides novel information regarding the network of areas implicated in the formation of automatic (letter–colour) associations. Synaesthesia-like binding of colours to letters by non-

Abstract from the 14th International Multisensory Research Forum, The Hebrew University of Jerusalem, Israel, 2013.
* Contact author: f.kusnir@psy.gla.ac.uk

 DOI:10.1163/22134808-000S0119

synaesthetes seems to be suppressed by dlPFC under normal conditions, and may be released by modulation of the dlPFC. We speculate that dlPFC-stimulation facilitates the formation of letter–colour binding, possibly by emphasizing relevant associations during task performance and/or by disinhibiting other brain areas (i.e., posterior parietal) involved in automatic, perceptual letter–colour binding.

Keywords
Synaesthesia, learning, tDCS

References

Cohen Kadosh and Iuculano (2011). *Abstract Hum. Brain Mapping.*
Esterman *et al.* (2006). *J. Cogn. Neurosci.* **18**, 1570–1576.
Kusnir and Thut (2012). *Neuropsychologia* **50**, 3641–3652.
Nikolić *et al.* (2007). *Psychol. Sci.* **18**, 481–486.

BRILL Multisensory Research 26 Supplement (2013) 162–163 brill.com/msr

Poster Presentation

Spatial imagery is more associated with unfamiliar than familiar haptic shape perception: Activation and connectivity analyses

Simon Lacey [1,*], **Randall Stilla** [1], **Karthik Sreenivasan** [2], **Gopikrishna Deshpande** [2] **and K. Sathian** [1,3]

[1] Department of Neurology, Emory University School of Medicine, Atlanta, GA, USA
[2] Department of Electrical & Computer Engineering, Auburn University, Auburn, AL, USA
[3] Departments of Rehabilitation Medicine & Psychology, Emory University School of Medicine & Rehabilitation R&D Center of Excellence, Atlanta, GA, USA

Abstract

We have proposed that haptic activation of the visually shape-selective lateral occipital complex (LOC) reflects a model of multisensory object representation in which the role of visual imagery is modulated by object familiarity (Lacey *et al.*, 2009). Supporting this, a visual object imagery task produced more correlations of activation magnitude and more similar patterns of effective connectivity with haptic perception of familiar (fHS) than unfamiliar (uHS) shapes (Deshpande *et al.*, 2010; Lacey *et al.*, 2010). Here, we hypothesize that visual spatial imagery (spIMG) would evoke activation and effective connectivity patterns more related to uHS than fHS. Participants ($n = 12$) performed spIMG, uHS and fHS tasks in two fMRI sessions. Conjunction analyses showed numerous parietal cortical foci common to both uHS and spIMG. Conjunction foci for fHS and spIMG involved multiple subcortical regions but also included some parietal foci. Activation magnitudes were correlated between spIMG and both fHs and uHS, with parietal foci featuring in each. The 2D correlation between the spIMG and uHS effective connectivity networks was significant and positive, including paths between the intraparietal sulcus (IPS) and bilateral LOC, as predicted by our model, and between IPS and primary somatosensory cortex. By contrast, the spIMG and fHS networks were non-significantly correlated. Thus, although the activation analyses suggested involvement of spatial imagery in haptic shape perception regardless of object familiarity, the effective connectivity analyses were consistent with our proposed model, supporting a stronger relation between spatial imagery and haptic perception of unfamiliar compared to familiar shapes.

Abstract from the 14th International Multisensory Research Forum, The Hebrew University of Jerusalem, Israel, 2013.

[*] Contact author: slacey@emory.edu

DOI:10.1163/22134808-000S0120

Keywords
Multisensory, lateral occipital complex, functional magnetic resonance imaging

References

Deshpande, G., Hu, X., Lacey, S., Stilla, R. and Sathian, K. (2010). Object familiarity modulates effective connectivity during haptic shape perception, *NeuroImage* **49**(3), 1991–2000.

Lacey, S., Tal, N., Amedi, A. and Sathian, K. (2009). A putative model of multisensory object representation, *Brain Topography* **21**(3-4), 269–274.

Lacey, S., Flueckiger, P., Stilla, R., Lava, M. and Sathian, K. (2010). Object familiarity modulates the relationship between visual object imagery and haptic shape perception, *NeuroImage* **49**(3), 1977–1990.

BRILL Multisensory Research 26 Supplement (2013) 164 brill.com/msr

Poster Presentation

A rhythmic deviation within a musical sequence induces neural activation in inferior parietal regions after short-term multisensory training

Claudia Lappe [1,*], **Olaf Steinsträter** [2] **and Christo Pantev** [1]

[1] Institute for Biomagnetism and Biosignalanalysis, University of Münster, Germany
[2] GSI Helmholzzentrum für Schwerionenforschung GmbH, Darmstadt, Germany

Abstract

The musical mismatch negativity in MEG represents the violation of a regularity in a musical sequence. It may be considered as a pre-attentive sensory change detection or as a top-down prediction error signal. Rhythmic and melodic deviations within a musical sequence elicit a stronger mismatch negativity in musically trained subjects than in novices indicating that acquired musical expertise leads to better discrimination accuracy of musical material and better predictions about upcoming musical events. Expectation violations to musical material could therefore recruit neural generators that reflect top-down processes that are based on musical knowledge.

To investigate the neural generators of musical deviance detection after musical training with rhythmic material we localized musical mismatch data from a previous MEG study with beamformer analysis. The training material focused on rhythmic progression and the MEG measurements on rhythmic violations. The beamformer analysis revealed neural activation bilaterally within the vicinity of auditory cortices and in the inferior parietal lobule, in an area that has recently been implied in temporal processing. We suggest that the musical training that subjects had received prior to the MEG measurement, has established an internal forward model linking a piano tone with a specific motor movement. This internal forward model might have supported and improved predictions about the timing of upcoming tones. Expectation violations were than presumably processed in the parietal lobule where timing and the motor system are closely linked.

Keywords

Mismatch negativity, music, beamforming, MEG

Abstract from the 14th International Multisensory Research Forum, The Hebrew University of Jerusalem, Israel, 2013.

* Contact author: clappe@uni-muenster.de

 DOI:10.1163/22134808-000S0121

BRILL Multisensory Research 26 Supplement (2013) 165 brill.com/msr

Poster Presentation

Leaky path integration is less leaky in full cue conditions

Markus Lappe * **and Britta Zepp**

Institute for Psychology, University of Muenster, Germany

Abstract

Humans typically underestimate travel distance during walking, which can be explained by leaky path integration. The amount of leak, i.e., the underestimation, is determined by the length of the path. Experiments both with blindfolded walking and visually-simulated walking in virtual reality have shown that for movements along curved paths that veer left and right around a central forward direction estimates of the start-to-end distance decrease as the path length increases. Leaky path integration for visual travel distance estimation thus takes place along the actually traversed path even when a straight beeline distance is calculated. We studied whether the same leaky path integration occurs during real self-motion in a full cue situation when visual, vestibular and proprioceptive cues are available together. Twenty subjects walked from a starting point to targets 20, 40 or 60 m away, guided by an experimenter. They walked either along a straight line, or along paths that deviated first to the right and then to the left (or *vice versa*) before they reached the end point. This increased the path length by 10%, 20% or 30%. Subjects then gave a verbal estimate of their beeline distance from the starting point. When compared to results of previous blindfolded and visually-simulated walking experiments, the full-cue estimates showed less dependence on path length, suggesting a smaller leak in path integration in the multisensory condition.

Keywords

Path integration, vision, vestibular, proprioception, motion

Abstract from the 14th International Multisensory Research Forum, The Hebrew University of Jerusalem, Israel, 2013.

* Contact author: mlappe@uni-muenster.de

 DOI:10.1163/22134808-000S0122

BRILL Multisensory Research 26 Supplement (2013) 166 brill.com/msr

Poster Presentation

Temporal expectancy selectively enhances audiovisual target detection

Fabrizio Leo *, **Rebecca Jacob, Karsta Benz, Miriam Kiehne and Tömme Noesselt**

Otto-von-Guericke-Universität, Germany

Abstract

We investigated the effect of attending to a specific period in time on low-intensity target detection in two psychophysical experiments. In the first experiment, participants were asked to detect a deviant target within a regular distractor sequence. Five experimental conditions were used: in the auditory-only condition, the target was a frequency modulated tone presented in a train of 14 pure tone distractors. In the visual-only condition, the target was a Gabor patch with a higher contrast compared to Gabor distractors. In all three AV conditions, auditory and visual trains of distractors were presented simultaneously but each condition differed for target type (i.e.: auditory-only, visual-only or AV target). In the second experiment, participants performed a frequency-discrimination instead of simple detection task. In both experiments, the target could appear either after a short or a long interval (400 or 1400 ms after first stimulus onset, respectively). We manipulated temporal expectancy (TE) by presenting trials in 'expect-early' blocks which consisted of 85% trials with early targets and 15% with late targets and *vice versa* for 'expect-late' blocks. We found that expecting a sensory event at a specific time point significantly enhances detection performance. This expectancy benefit is particularly pronounced for the AV conditions in the expect-early compared to late condition. In the discrimination experiment, a significant TE benefit was observed only in the AV condition with AV target. Collectively, our data extend previous findings from unisensory expectancy and demonstrate a selective TE effect being more pronounced for AV compared to unisensory conditions.

Keywords

Temporal expectancy, audiovisual, integration, psychophysics

Abstract from the 14th International Multisensory Research Forum, The Hebrew University of Jerusalem, Israel, 2013.

* Contact author: fleo@med.ovgu.de

 DOI:10.1163/22134808-000S0123

BRILL Multisensory Research 26 Supplement (2013) 167 brill.com/msr

Poster Presentation

Synchrony perception between looming sounds and static touches: A MEG study on audiotactile integration

Elisa Leonardelli [1,*], **Christoph Braun** [1,2], **Nathan Weisz** [1,3], **Valeria Occelli** [4,5] and **Massimiliano Zampini** [1,4]

[1] Center for Mind/Brain Sciences, University of Trento, Rovereto, Italy
[2] MEG Center, University of Tübingen, Germany
[3] Department of Psychology, University of Konstanz, Konstanz, Germany
[4] Department of Psychology and Cognitive Sciences, University of Trento, Rovereto, Italy
[5] Department of Neurology, Emory University, Atlanta, GA, USA

Abstract

Aim of this study was to investigate the neural correlates of audio and tactile integration by means of magnetoenceohalography (MEG). Participants were presented with a one-second sound of a mosquito approaching the head from the left side. Furthermore a spatially correlated tactile stimulus was presented at different stimulus onset asynchronies (SOAs), yielding ambiguous perceptions as to whether the touch co-occured simultaneously with the end of the sound. Participants were asked to evaluate whether the approaching mosquito and the touch belonged to the same event (i.e., 'integrated': the impression that it was the approaching mosquito that touched them) or not (i.e., 'segregated').

In a first run, we calculated the temporal thresholds (difference of offset sound to onset touch) at which the response's rate integration/segregation was at 50% by means of a staircase procedure. Interestingly participants showed a larger temporal window of integration for tactile stimuli presented before the end of the sound.

In the subsequent MEG experiment, data were recorded for conditions with SOAs individually set in order to match the 50% rate of integration. In this way, bimodal physically identical stimuli that elicit different multisensory perceptions, have been directly compared for avoiding possible confounds introduced by linear additive model.

A comparison between trials in which participants reported an integrated percept *versus* a segregated percept, showed a decrease in the alpha power in the 350 ms time window preceding the tactile stimulus, for the condition were integration was reported. This pattern supports the influence of ongoing oscillatory brain activity on perception of integration between different senses.

Keywords

Audiotactile, looming, MEG, oscillations

Abstract from the 14th International Multisensory Research Forum, The Hebrew University of Jerusalem, Israel, 2013.

* Contact author: elisa.leonardelli@unitn.it

DOI:10.1163/22134808-000S0124

BRILL Multisensory Research 26 Supplement (2013) 168 brill.com/msr

Poster Presentation

Dissociating sensory attention and motor preparation: An ERP study

Pia Ley * **and Brigitte Röder**

Biological Psychology and Neuropsychology, University of Hamburg, Germany

Abstract

Sensory and motor systems are closely linked as expressed in the Premotor Theory of Attention (PMTA, Rizzolatti *et al.*, 1987). The PMTA predicts that sensory attention effects on stimulus processing are equivalent to an explicit or implicit motor preparation towards these stimuli. This view has been supported by findings suggesting a high overlap of neural systems involved in attention control and motor preparation (Corbetta *et al.*, 1998).

However, attempts to dissociate sensory attention and motor preparation effects, which, if possible, would argue against the PMTA, are still lacking. Eleven participants (8 female, mean age: 24 years (20–32 years)) performed a dual task: First they were verbally instructed to prepare a unimanual movement either towards the center, the left or the right side (motor attention). During the movement preparation phase a stream of visual stimuli was presented in the left and the right hemifield. Keeping central fixation participants had to detect rare color changes in one hemifield (left or right), while ignoring all other stimuli. After a variable interval (1–10 s) an auditory stimulus was presented that instructed the participants to execute the movement. This experimental paradigm allowed us to dissociate sensory and motor attention effects on visual stimulus processing.

We analyzed the behavioral response to target stimuli and the visually ERPs to irrelevant visual stimuli at frontal, central and occipital cluster sites at three different time windows centered over the P1, the N1 and the N2. Both effects of sensory and motor attention as well as the interaction were analyzed.

Keywords

Sensory attention, movement preparation, premotor theory of attention, ERP

References

Corbetta, M., Akbudak, E., Conturo, T. E., Snyder, A. Z., Ollinger, J. M., Drury, H. A., Linenweber, M. R., Petersen, S. E., Raichle, M. E., Van Essen, D. C. and Shulman, G. L. (1998). A common network of functional areas for attention and eye movements, *Neuron* **21**(4), 761–773.

Rizzolatti, G., Riggio, L., Dascola, I. and Umiltà, C. (1987). Reorienting attention across the horizontal and vertical meridians — evidence in favor of a preomotor theory of attention, *Neuropsychologia* **25**(1), 31–40.

Abstract from the 14th International Multisensory Research Forum, The Hebrew University of Jerusalem, Israel, 2013.

* Contact author: pia.ley@uni-hamburg.de

DOI:10.1163/22134808-000S0125

BRILL Multisensory Research 26 Supplement (2013) 169 brill.com/msr

Poster Presentation

Does metal always sound stiffer than drum?

Juan Liu * **and Hiroshi Ando**

Multisensory Cognition and Computation Laboratory,
Universal Communication Research Institute, NICT, Japan

Abstract

In our former work it has been shown that contact sounds of different materials influence the haptic perception of stiffness and surfaces with metal sound gave people the impression of a harder surface than drum sound. Because sound parameters (i.e., frequency and damping parameters) of various materials are quite different, it is not clear whether the higher level knowledge of material or the low level sound parameters have stronger effects. Here we modified the two original sounds of metal and drum by adjusting their mode frequency centroids to be the same while keeping the internal friction coefficient (i.e., the multiply of frequency and damping parameter) constant. Thus in the interview all participants still recognized the modified sounds as metal and drum and subjectively evaluated surfaces with metal sounds were stiffer than those with drum sounds. However in the task of adjusting stiffness of surfaces to match the memorized stiffness of standard surfaces with different sounds, statistical analysis showed interaction effect of material and parameter factors. Although the perceived stiffness of surfaces with original metal and drum sounds were significantly different, surface with modified metal sound was perceived no stiffer than that with modified drum sound. It suggests that sound parameters may influence the haptic-auditory encoding of stiffness stronger than the knowledge of materials.

Keywords

Haptic-auditory integration, stiffness perception, material, contact sound

---

Abstract from the 14th International Multisensory Research Forum, The Hebrew University of Jerusalem, Israel, 2013.

* Contact author: juanliu@nict.go.jp

 DOI:10.1163/22134808-000S0126

BRILL Multisensory Research 26 Supplement (2013) 170 brill.com/msr

Poster Presentation

Auditory and tactile signals combine to influence vision during binocular rivalry

Claudia Lunghi [1,2,*], Maria C. Morrone [3,4], David C. Burr [1,2] and David Alais [5]

[1] Department of Neuroscience, Università Degli Studi di Firenze, Italy
[2] Institute of Neuroscience, CNR — Pisa, Italy
[3] Department of Translational Research on New Technologies in Medicine and Surgery, Università di Pisa, Italy
[4] Scientific Institute Stella Maris, Pisa, Italy
[5] School of Psychology, University of Sydney, Australia

Abstract

Resolution of perceptual ambiguity is a major function of cross-modal interactions, making the study of bistable perception in multisensory contexts a powerful and revealing tool. We previously used binocular rivalry, a visual bistable phenomenon, to show that touch can specifically interact with vision to resolve spatial conflict between the eyes (Lunghi, Binda and Morrone, 2010). Here we investigate whether auditory and tactile stimuli can influence binocular rivalry generated by interocular temporal conflict. Using visual stimuli of different temporal frequencies (spatio-temporal noise filtered at 3.75 or at 15 Hz) to produce visual perceptual alternations, we added an amplitude modulated sound or a vibration that was congruent with one or the other visual temporal frequencies. We found that auditory and tactile stimulation interacted with binocular rivalry by promoting dominance of the congruent visual stimulus. This effect depended on the strength of the auditory/tactile stimulus, and was absent when a modulation depth declined to 33%. However, when auditory and tactile stimuli that were too weak on their own to bias binocular rivalry were combined, their influence over vision was very strong, suggesting the auditory and tactile temporal signals were summated. When auditory and tactile stimuli were presented at maximum strength, but temporally in anti-phase, they had no influence over vision, a null effect that again suggests audio-tactile summation in the temporal domain. These results indicate a functional link between auditory and tactile low temporal frequency channels, suggesting the existence of common neural substrates for the two sensory modalities.

Keywords

Cross-modal processing, binocular rivalry, psychophysics

Abstract from the 14th International Multisensory Research Forum, The Hebrew University of Jerusalem, Israel, 2013.

* Contact author: claudia.lunghi@in.cnr.it

 DOI:10.1163/22134808-000S0127

BRILL Multisensory Research 26 Supplement (2013) 171 brill.com/msr

Poster Presentation

Blind in a virtual world — using distance information to accomplish virtual tasks

Shachar Maidenbaum [1,*], Shelly Levy-Tzedek [1,2], Daniel-Robert Chebat [1,2] and Amir Amedi [1,2,3]

[1] Department of Medical Neurobiology, The Institute for Medical Research Israel–Canada (IMRIC), Faculty of Medicine, The Hebrew University of Jerusalem, Israel
[2] The Edmond and Lily Safra Center for Brain Sciences (ELSC), The Hebrew University of Jerusalem, Israel
[3] The Cognitive Science Program, The Hebrew University of Jerusalem, Jerusalem, Israel

Abstract

Distance information is critical to our understanding of our surrounding environment, especially in virtual reality settings. Unfortunately, as we gage distance mainly visually, the blind are prevented from properly utilizing this parameter to formulate 3D cognitive maps and cognitive imagery of their surroundings. Our purpose is to increase the accessibility of virtual environments to the blind using distance information which they will receive as auditory information. We aim to create a setup which will enable the blind and visually impaired to experience novel environments virtually before travelling to them in the real world and additionally will enable the blind better accessibility to virtual environments for purposes such as entertainment and education.

Here, blind and sighted-blindfolded subjects performed navigation and shape-discrimination tasks in virtual environments, using a simple transformation between virtual distance and sound, based on the concept of a virtual guide cane (paralleling in a virtual environment the 'EyeCane', developed in our lab).

We show qualitatively that with minimal training it is possible for blind and blindfolded subjects to easily learn this transformation, enabling the discrimination of virtual 3D orientation and shapes and navigation in basic virtual environments using a standard mouse and audio-system.

Keywords

Sensory substitution, blind, navigation, virtual environment

Reference

Maidenbaum, S., Arbel, R., Abboud, S., Chebat, D. R., Levy-Tzedek, S. and Amedi, A. (2012). A virtual 3D shape and orientation discrimination using point distance information, in: *Proc. of the 9th Intl Conf. Disability, Virtual Reality & Associated Technologies*.

Abstract from the 14th International Multisensory Research Forum, The Hebrew University of Jerusalem, Israel, 2013.

* Contact author: Shachar.Maidenbaum@mail.huji.ac.il

 DOI:10.1163/22134808-000S0128

BRILL Multisensory Research 26 Supplement (2013) 172 brill.com/msr

Poster Presentation

Visual and vestibular orientation cues interact to influence perceived depth

Charles M. Mander * **and Laurence R. Harris**

York University, Canada

Abstract

Models of depth perception typically omit orientation despite the potential usefulness of the ground plane and of head position information for interpreting disparity (Blohm *et al.*, 2008). Perceived depth is shortened when visual cues to upright are rotated relative to gravity (Mander and Harris, 2012), suggesting a role of orientation in depth perception. Here we used the York University Tumbled and Tumbling Room facilities, realistically decorated rooms systematically arranged to vary the relative orientation of visual, gravity and body cues to upright. We exploited size/distance constancy to assess perceived depth. Observers matched the perceived length of a tactile rod to a visual line (controlled by a QUEST adaptive procedure) projected on the wall of the facilities.

The line was consistently set longer when the room and subject were rotated 90° relative to gravity, compared to when subjects were upright suggesting that the opposite wall appeared closer. The effect was modulated by binocular cues, nearly doubling under monocular viewing. After systematically varying the room's and subject's orientation, the largest 'expansion' was obtained when observers were upright in a rotated room (looking at the ceiling), suggesting that the illusion is induced primarily by rotation of visual cues. The effect was reduced when subjects were rotated toward the ceiling in an upright room, suggesting that the perceived distance change induced by perceived body tilt is modulated by actual body tilt. Thus, vestibular orientation cues are a mediating factor when there is a mismatch between visual and gravitational upright.

Keywords

Psychophysics, depth perception, distance perception, illusion

References

Blohm, G., Khan, A. Z., Ren, L., Schreiber, K. M. and Crawford, J. D. (2008). Depth estimation from retinal disparity requires eye and head orientation signals, *J. Vis.* **8**(16), 3.1-3.23.

Mander, C. M. and Harris, L. R. (2012). An interaction between body orientation and gravity produces a novel illusion in visual distance perception, *Seeing & Perceiving* **25**(supplement), 116.

Abstract from the 14th International Multisensory Research Forum, The Hebrew University of Jerusalem, Israel, 2013.

* Contact author: cmander@yorku.ca

 DOI:10.1163/22134808-000S0129

BRILL Multisensory Research 26 Supplement (2013) 173–174 brill.com/msr

Poster Presentation

On the colours of odours: Are the French really so different from the British?

Yelena Maric [1,*], **Carlos Velasco** [2], **Charles Spence** [2] **and Muriel Jacquot** [1]

[1] LIBio, Université de Lorraine, France

[2] Crossmodal Research Laboratory, Department of Experimental Psychology, University of Oxford, UK

Abstract

It has been argued that culture-specific experiences exert a profound influence on various basic aspects of odour perception. Differences in terms of odour-colour associations might therefore be expected as a function of the culture in which people live, especially when the functions of odours and the symbolism of colours differ between countries (Maric and Jacquot, 2012). In the present study, we investigated the influence of experience on odour-colour correspondences at a European level.

In Experiment 1, the responses of 59 British and 60 French participants to the same set of 16 odorants were compared. In each country, untrained participants were first presented with odorants and asked to select among 24 colours the one that best matched each given olfactory stimulus. Next, they rated each odour according to 4 descriptors: intensity, familiarity, pleasantness and edibility. Significant colour characterizations were observed for all the tested odours in both populations. No differences were reported between French and British groups in terms of the odour-colour associations that they exhibited.

In Experiment 2, 'chromatic cards', generated for each odorant on the basis of the colours associated by the French participants were generated. 60 British participants were presented with 3 chromatic cards and asked, first, to name the odour which each card evocated, and second, to select from amongst 3 different olfactory stimuli the one that best matched each card. The results revealed that French chromatic cards were significantly associated to their related odours by the British participants.

These results highlight the existence of robust crossmodal odour-colour correspondences among Northern Europeans. Our results raise some important questions concerning the representation of odours in different cultures. The practical applications of our findings will also be discussed.

Abstract from the 14th International Multisensory Research Forum, The Hebrew University of Jerusalem, Israel, 2013.

* Contact author: yelena.maric@univ-lorraine.fr

DOI:10.1163/22134808-000S0130

Keywords
Crossmodal, cross-cultural, odour-colour

References

Maric, Y. and Jacquot, M. (2012). Contribution to understanding odour-colour associations, *Food Quality and Preference*. Available online: http://dx.doi.org/10.1016/j.foodqual.2012.05.001.

Maric, Y., Barbar, R. and Jacquot, M. (2012). As pink as an orange blossom odour: a French–Lebanese cross-cultural study, in: *Proceedings of AIC Color 2012, Interim Meeting of the International Color Association*.

BRILL Multisensory Research 26 Supplement (2013) 175 brill.com/msr

Poster Presentation

Perceptual training alters the time window of multisensory integration

David P. McGovern [1,*], **Neil W. Roach** [2], **Eugenie Roudaia** [1] **and Fiona N. Newell** [1]

[1] Trinity College Institute of Neuroscience, Trinity College Dublin, Ireland

[2] Visual Neuroscience Group, The University of Nottingham, UK

Abstract

Given the inherently noisy nature of sensory signals, the brain must maintain a degree of tolerance for temporal discrepancies when combining multisensory information. While the limits of this temporal window of integration are well established, little is known about how they are affected by sensory experience. Powers *et al.* (2009) reported that training on an audiovisual simultaneity task narrows the temporal window of integration. However, because subjects were trained on the same task that was used to measure their integration windows, it is unclear whether this effect reflects a genuine change in multisensory integration or a simple improvement in task performance. Here, we examine whether perceptual training affects multisensory integration using independent training and test measures. In pre- and post-training sessions, participants were required to report the position of an auditory noise burst while ignoring a visual stimulus, across a range of stimulus onset asynchronies. The ventriloquist effect induced by the visual stimulus was greatest for simultaneous stimuli and declined with increasing inter-stimulus asynchrony. Training consisted of three 45-minute sessions where participants practiced a 2-IFC audiovisual simultaneity task. Training resulted in narrowing of the time window of integration comparable to that observed by Powers *et al.*, as well as a reduction in the amplitude of the peak ventriloquist effect. Unisensory localisation thresholds were unaffected by training, ruling out changes in auditory and visual sensitivity as the cause of improvements. These findings demonstrate that perceptual training can affect multisensory integration and have implications for the future development of multisensory training protocols.

Keywords

Multisensory integration, perceptual learning, temporal processing

Reference

Powers, A. R., III, Hillock, A. R. and Wallace, M. T. (2009). Perceptual training narrows the temporal window of multisensory binding, *J. Neurosci.* **29**, 12265–12274.

Abstract from the 14th International Multisensory Research Forum, The Hebrew University of Jerusalem, Israel, 2013.

* Contact author: mcgoved1@tcd.ie

DOI:10.1163/22134808-000S0131

BRILL Multisensory Research 26 Supplement (2013) 176–177 brill.com/msr

Poster Presentation

Use of a system to provide independent views to both eyes with a function to generate artificial saccade movements

Fumio Mizuno [1,*], **Tomoaki Hayasaka** [2] **and Takami Yamaguchi** [2]

[1] Tohoku Institute of Technology, Japan

[2] Tohoku University, Japan

Abstract

In previous work, we developed a wearable system that was named Virtual Chameleon that gives a user an artificial ability to move both eyeballs independently (Mizuno *et al.*, 2009, 2010). The users could look around, discern between right and left views while independent views were given to both eyes, though binocular rivalry originating from use of Virtual Chameleon affected the vision of the users (Mizuno *et al.*, 2011).

The most eye movements performed by humans are saccadic movement but a control method of actuators equipped with Virtual Chameleon was corresponding to smooth eye movements. Therefore, we focused on influence of saccadic eye movements on visual perception with binocular rivalry, and implemented functions to produce artificial saccadic movements in Virtual Chameleon. In this research, we performed fundamental experiments on seven volunteers to evaluate effects on implementation of artificial saccadic eye movements. The experiments were tests to find out each position of targets put in both side of a subject. As experimental results, accuracy rates and response times to find out target positions were obtained. Accuracy rates were from 0.88 to 1.0, and averages of response time were from 2.41 ± 0.20 [s] to 4.34 ± 0.84 [s]. It was indicated that user can look around and distinguish independent views as previous wok and the implementation of the functions to generate artificial saccade movements to Virtual Chameleon improved the delay of up to 5% on five subjects compared to previous system. It was suggested that the function to switch control timing of the system and generate saccadic eye movements improve human performance with use of Virtual Chameleon.

Keywords

Binocular rivalry, saccade eye movements, virtual reality

Abstract from the 14th International Multisensory Research Forum, The Hebrew University of Jerusalem, Israel, 2013.

* Contact author: fumio@tohtech.ac.jp

DOI:10.1163/22134808-000S0132

References

Mizuno, F., Hayasaka, T. and Yamaguchi, T. (2009). Virtual Chameleon — a system to provide different views to both eyes, in: *IFMBE Proceedings*, Vol. 25, pp. 169–172.

Mizuno, F., Hayasaka, T. and Yamaguchi, T. (2010). A portable device to represent different views to both eyes, in: *Proc. the 32nd Annual International Conference of the IEEE EMBS*, pp. 3210–3213.

Mizuno, F., Hayasaka, T. and Yamaguchi, T. (2011). A fundamental evaluation of human performance with use of a device to present different two-eyesight both eyes, in: *5th European Conference of International Federation Medical and Biological Engineering 2011, Proc. of IFMBE*, Vol. 37, pp. 1176–1179.

BRILL Multisensory Research 26 Supplement (2013) 178 brill.com/msr

Poster Presentation

Audition dominates vision after loss of one eye early in life when the systems are in competition but not when they are integrated

Stefania S. Moro [1,2,*], **Laurence R. Harris** [1] **and Jennifer K. E. Steeves** [1,2]

[1] Centre for Vision Research, York University, Canada

[2] The Hospital for Sick Children, Canada

Abstract

Previous research has shown that people with one eye have enhanced sound localization and lack visual dominance, commonly found in binocular and monocular viewing controls. These findings imply cross-sensory adaptation likely as compensation for their loss of binocularity. We assessed whether the advantage given to audition in people with one eye, when the auditory and visual systems were in competition, might also be found when the systems were integrated together to make unified judgements. Participants were asked to spatially localize perceptually fused audiovisual events in which the auditory and visual components were spatially disparate in order to quantify the relative weightings assigned to each system when the systems were integrated. There was no difference in the reliability assigned to localizing unimodal visual and auditory targets by people with one eye compared to controls. When localizing bimodal targets, the weightings assigned to each sensory modality in both people with one eye and controls were predictable from their unimodal performance in accordance with the Maximum Likelihood Estimation (MLE) model. People with one eye appear to integrate the auditory and visual components of multisensory events optimally when determining spatial location despite the fact that they do not show the typical dominance of vision over audition when the two systems are in competition. It is possible that attentional modifications to the processing of each component when they are processed in parallel may represent an adaptive cross-sensory compensatory mechanism for the loss of binocular visual input that does not alter how these signals are integrated.

Keywords

Auditory, visual, multisensory, monocular, binocular

Abstract from the 14th International Multisensory Research Forum, The Hebrew University of Jerusalem, Israel, 2013.

* Contact author: smoro@yorku.ca

DOI:10.1163/22134808-000S0133

BRILL Multisensory Research 26 Supplement (2013) 179 brill.com/msr

MULTISENSORY RESEARCH

Poster Presentation

Effects of audiovisual motion coherence on multisensory processing

Aaron Nidiffer *

Wallace Laboratory, Vanderbilt University, Nashville, TN, USA

Abstract

Frequently, the presentation of stimuli from two sensory modalities results in behavioral benefits, including improvements in the detection and accuracy of responses and a speeding of response times. Although these benefits have been shown using both simple and complex multisensory stimulus pairs, they have not been extended to the domain of motion coherence. We developed auditory motion stimuli to parallel the varying motion coherence levels of random dot kinematograms (RDKs) by simulating the motion of a broadband noise within a static uncorrelated broadband noise. Using these auditory stimuli and visual RDKs at varying levels of motion coherence, we tested the behavioral gains associated with multisensory pairs. Both the accuracy and timing of responses to unisensory (i.e., auditory alone, visual alone) stimuli versus multisensory stimuli were assessed at four different levels of motion coherence: 10%, each individual's threshold, average threshold for our population, 70%. In addition, responses were assessed for congruently or incongruently moving stimuli (i.e., the auditory and visual motion either moved in the same direction or opposite directions). We found that maximum gains in multisensory performance (e.g., greater response accuracy and more race-model violations) were seen at threshold levels of motion coherence; levels at which the unisensory cues have maximum ambiguity. This study further reinforces the utility of multisensory integration for weakly effective stimuli, and expands the multisensory toolbox to include motion stimuli used in the examination of sensory processing in clinical populations including autism and schizophrenia.

Keywords

Multisensory, motion perception, inverse effectiveness, random dot kinematogram

Abstract from the 14th International Multisensory Research Forum, The Hebrew University of Jerusalem, Israel, 2013.

* Contact author: aaron.r.nidiffer@vanderbilt.edu

 DOI:10.1163/22134808-000S0134

BRILL Multisensory Research 26 Supplement (2013) 180 brill.com/msr

Poster Presentation

Effects of combining visual and olfactory information on affective impressions

Yurie Nishino *, **Dong Wook Kim and Hiroshi Ando**

Multisensoy Cognition and Computation Laboratory, Universal Communication Research Institute, National Institute of Information and Communications Technology (NICT), Japan

Abstract

We receive different impressions from different visual images or odors, but it was not clear what kinds of impression the combination of visual images and odors elicits. Thus, we conducted experiments where we compared affective impressions when only olfactory information, only visual information, and combinations of olfactory and visual information were presented. In this experiment, the subjects were asked to rate the impression in 'mild-sharp' and 'clear-dense' dimensions using a visual-analogue-scale method when they were shown each of 8 different odors, 8 different visual images, or combinations of these different types of visual images and odors. The subjects were also asked to judge how much each specific combination of odor and visual image was harmonious on a 7-point scale. The results show some averaging effects when visual and olfactory information fell into different categories (i.e., mild, sharp, clear or dense) in the impression space. On the other hand, the impression of image–odor combination tended to be enhanced when the impression of odor or image alone fell into the same category along with the condition when they were judged as being harmonious. These results suggest that the way combination of visual and olfactory information elicits the impression may depend on the impression of individual visual and olfactory information and whether they are harmonious.

Keywords

Odor, visual image, combination, impression

Abstract from the 14th International Multisensory Research Forum, The Hebrew University of Jerusalem, Israel, 2013.

* Contact author: ynishino@nict.go.jp

DOI:10.1163/22134808-000S0135

BRILL Multisensory Research 26 Supplement (2013) 181 brill.com/msr

Poster Presentation

Audio-tactile multisensory integration as a proxy for full body peri-personal space

Jean-Paul Noel [1,*], **Giulia Galli** [2], **Olaf Blanke** [1] **and Andrea Serino** [3]

[1] Ecole Polytechnique Fédérale de Lausanne, Switzerland
[2] University of California, Davis, USA
[3] Università di Bologna, Italy

Abstract

Peri-personal space (PPS) — i.e., the space immediately surrounding the body — has been traditionally conceived as the space reachable through limb movements, and therefore, it remained centered on a specific peripheral extremity (most prominently, the hand and the face). However, we do experience our body in its space as a whole and not only as a discrete addition of body parts, thus begetting the question: Is there a full body representation of PPS? And if so, what are its characteristics and boundaries? In a first set of experiments, through the use of a novel audio-tactile paradigm, we (i) established the presence of a full body PPS, (ii) determined its boundary both in the front and backspace, and (iii) compared its spatial extent to the one of the peri-hand space. In addition, we observed a meshing between peri-trunk and peri-hand space when these two are in close physical proximity, and established a direct relationship, as it's prominently shown in monkey electrophysiology data, between the velocity of incoming stimuli and the expansion of PPS. Our findings argue for the possibility of utilizing an audio-tactile multisensory integration paradigm in order to measure full body PPS, as well as establish the aforementioned space characteristics. We show analogous results to those observed in electrophysiology data, as well as expand on previous literature by showing an interplay between different conceptions of PPS.

Abstract from the 14th International Multisensory Research Forum, The Hebrew University of Jerusalem, Israel, 2013.

* Contact author: jeanpaulnc@gmail.com

 DOI:10.1163/22134808-000S0136

BRILL Multisensory Research 26 Supplement (2013) 182–183 brill.com/msr

MULTISENSORY RESEARCH

Poster Presentation

Bold responses to tactile stimuli in visual and auditory cortex depend on the frequency content of stimulation

Per F. Nordmark [1,2,3,*], **J. Andrew Pruszynski** [1,2] **and Roland S. Johansson** [1,2]

[1] Umeå Center for Functional Brain Imaging, Umeå University, Sweden

[2] Dept. of Integrative Medical Biology, Physiology Section, Umeå University, Sweden

[3] Dept. of Surgical and Perioperative Sciences, Umeå University, Sweden

Abstract

Although some brain areas preferentially process information from a particular sensory modality, these areas can also respond to other modalities (Amedi *et al.*, 2001; Ghazanfar and Schroeder, 2006; Macaluso and Driver, 2005; Merabet and Pascual-Leone, 2010; Schroeder *et al.*, 2001). We used fMRI to show that such responsiveness to tactile stimuli depends on the temporal frequency of stimulation. Participants performed a tactile threshold-tracking task where the tip of either their left or right middle finger was stimulated at 3, 20, or 100 Hz. Whole-brain analysis revealed an effect of stimulus frequency in two regions: the auditory cortex and the visual cortex. The BOLD response in the auditory cortex was stronger during tactile stimulation at hearable frequencies (20 and 100 Hz) whereas the response in the visual cortex was suppressed at infrasonic frequencies typical of biological motion (3 Hz). Regardless of which hand was stimulated, the frequency dependent effects were lateralized to the left auditory cortex and the right visual cortex. Furthermore, the frequency-dependent effects in both areas were abolished when the participants performed a visual task while receiving identical tactile stimulation as in the tactile threshold-tracking task. Our findings advance the metamodal theory of brain function (Pascual-Leone and Hamilton, 2001) by showing that changing input to a single sensory modality in the context of the same task can affect BOLD activity in two unimodal cortical areas primarily associated with processing other sensory modalities. That is, brain areas contribute to sensory processing by performing specific computations regardless of input modality.

Keywords

Human, touch, auditory cortex, visual cortex, somatosensory cortex

Abstract from the 14th International Multisensory Research Forum, The Hebrew University of Jerusalem, Israel, 2013.

* Contact author: per.nordmark@physiol.umu.se

DOI:10.1163/22134808-000S0137

References

Amedi, A., Malach, R., Hendler, T., Peled, S. and Zohary, E. (2001). Visuo-haptic object-related activation in the ventral visual pathway, *Nature Neuroscience* **4**, 324–330.

Ghazanfar, A. A. and Schroeder, C. E. (2006). Is neocortex essentially multisensory? *Trends in Cognitive Sciences* **10**, 278–285.

Macaluso, E. and Driver, J. (2005). Multisensory spatial interactions: a window onto functional integration in the human brain, *Trends in Neurosciences* **28**, 264–271.

Merabet, L. B. and Pascual-Leone, A. (2010). Neural reorganization following sensory loss: the opportunity of change, *Nature Reviews Neuroscience* **11**, 44–52.

Schroeder, C. E., Lindsley, R. W., Specht, C., Marcovici, A., Smiley, J. F. and Javitt, D. C. (2001). Somatosensory input to auditory association cortex in the macaque monkey, *Journal of Neurophysiology* **85**, 1322–1327.

Pascual-Leone, A. and Hamilton, R. (2001). The metamodal organization of the brain, *Progress in Brain Research* **134**, 427–445.

BRILL Multisensory Research 26 Supplement (2013) 184–185 brill.com/msr

Poster Presentation

The influence of gender incongruence on the McGurk-percept: A combined behavioural and fMRI study

Annika Notbohm [1,*], Marcus J. Naumer [1], Jasper J. F. van den Bosch [1], Jochen Kaiser [2] and Jason S. Chan [1]

[1] Crossmodal Object Imaging Lab, Institute of Medical Psychology, Goethe-University Frankfurt am Main, Germany

[2] Institute of Medical Psychology, Goethe-University Frankfurt am Main, Germany

Abstract

The McGurk-effect (McGurk and MacDonald, 1976) is a robust illusion which is broadly studied in the context of audiovisual integration. In the illusion, auditory speech perception is modified by discrepant visual lip-movements when presented synchronously, leading to a third, not physically present percept. There is an ongoing debate as to whether the percept is integrated early in the sensory system or if it is susceptible to cognitive intervention. Here, the McGurk-effect was studied using gender congruency (face/voice stimuli) as the cognitive intervention to address the question of integration. We first investigated changes in prevalence and reaction times due to perceived gender discrepancy during the McGurk-percept. In the second experiment, neural correlates of the gender discrepant McGurk-percept were studied using fMRI.

We did not find differences in reaction times or accuracy between gender congruent or incongruent stimuli. Thus it appears the McGurk-percept itself is unaffected by gender incongruent stimuli on a behavioural level.

In the fMRI experiment, we found the superior temporal gyrus to reveal a significantly increased activity when the McGurk-effect is perceived as compared to non-McGurk trials, but no effect in this area was found for gender congruency. However, a 2×2 whole brain ANOVA revealed a significant interaction in the face identity processing area (fusiform gyrus) as well as inferior parietal gyrus and superior colliculus. We suggest a mechanism, allowing a stable illusory percept in the gender discrepant McGurk-percept to be facilitated. To our knowledge this is the first study to demonstrate on a neural level that gender congruency affects processing of the McGurk-effect.

Abstract from the 14th International Multisensory Research Forum, The Hebrew University of Jerusalem, Israel, 2013.

* Contact author: anotboh@googlemail.com

 DOI:10.1163/22134808-000S0138

Keywords
McGurk-effect, fMRI, semantic congruency, gender, audiovisual integration

Reference

McGurk, H. and MacDonald, J. (1976). Hearing lips and seeing voices, *Nature* **264**(5588), 746–748.

BRILL Multisensory Research 26 Supplement (2013) 186 brill.com/msr

MULTISENSORY RESEARCH

Poster Presentation

Spatial imagery and cognitive style in blind individuals

Valeria Occelli [1,*], **Jonathan B. Lin** [1], **Simon Lacey** [1] **and K. Sathian** [1,2]

[1] Department of Neurology, Emory University, Atlanta, GA, USA

[2] Departments of Rehabilitation Medicine and Psychology, Emory University and Rehabilitation R&D Center of Excellence, Atlanta, GA, USA

Abstract

Previous research into the effects of blindness on spatial imagery has been inconclusive. Here we investigated whether visual status and cognitive style affected performance on a spatial imagery task (spIMG). Early blind, late blind and age-matched sighted participants were recruited. Blind participants were classified as early or late blind according to the age at which form vision was lost (cutoff = 10 years). Participants were asked to memorize a 4 × 4 lettered matrix. Subsequently, when cued auditorily by four-letter sequences, they mentally constructed patterns within the matrix and compared the resulting shapes. They also completed the Object-Spatial Imagery and Verbal Questionnaire (OSIVQ; Blazhenkova and Kozhevnikov, 2009). Overall, the spIMG accuracy of both early blind and late blind participants was significantly poorer than that of sighted participants. Moreover, the accuracy of blind participants was uncorrelated with the age at which form vision was lost. Imagery preference, as assessed by the OSIVQ, varied with the visual status of the participants, with early (but not late) blind participants reporting significantly lower object scores than sighted controls. Cognitive style also differently modulated performance across the participant groups: higher scores for either object imagery or verbalizing were significantly negatively correlated with spIMG accuracy of sighted, but not blind participants, whereas higher spatial imagery scores were significantly positively correlated with spIMG accuracy only in the early blind group. These findings suggest the potential for spatial cognitive training of blind people, and also imply the need for individualizing such training based on cognitive style.

Keywords

Spatial imagery, blindness

Reference

Blazhenkova, O. and Kozhevnikov, M. (2009). The new object-spatial-verbal cognitive style model: Theory and measurement, *Applied Cognitive Psychology* **23**(5), 638–663.

Abstract from the 14th International Multisensory Research Forum, The Hebrew University of Jerusalem, Israel, 2013.

* Contact author: valeria.occelli@emory.edu

DOI:10.1163/22134808-000S0139

BRILL Multisensory Research 26 Supplement (2013) 187 brill.com/msr

Poster Presentation

Neural correlates of audio-visual perception in 3D-surround cinematography

Akitoshi Ogawa *, **Cécile Bordier and Emiliano Macaluso**

Santa Lucia Foundation, Italy

Abstract

Recent technological developments have made popular home-based systems to present visual stimuli in 3D and sounds from multiple sources. These systems generate a rich and multisensory percept of external space, with an augmented sense of scene-structure and enhanced sound-localization. We investigated the neural correlates of 3D-surround cinematography using stereoscopy and a multi-speakers system during fMRI. Subjects watched a commercial 3D movie presented in four, factorially-organized conditions: 3D-surround, 3D-mono, 2D-surround or 2D-mono (blocks of 26–57 s). They were then asked to evaluate the 'sense of presence' of short clips extracted from the movie, again presented in the 4 conditions. The imaging results showed a main effect of 3D viewing in LOC, V3A and SPL. The main effect of surround sound activated the auditory cortex, while an interaction between vision and audition was observed in the MTG, inferior–anterior to the auditory cortex. In MTG maximal activation was found in the 2D-surrond condition. This region may perform the analysis of complex auditory scenes, and more so in the absence of redundant visuo-spatial depth information. The appraisal results revealed that 3D-vision was the main factor triggering an enhanced sense of presence. Consistent with this, the presence scores were associated with activity in lateral and dorsal occipital areas, but also with visual scene-related areas in the ventral occipito-temporal cortex. These data show that the perception of 3D-surround stimuli relies on an extensive set of visual and auditory areas, implicating these regions in the processing of the complex multisensory signals that characterize our everyday-life experience.

Keywords

fMRI, cinematography, stereoscopic depth, surround sound, sense of presence

---

Abstract from the 14th International Multisensory Research Forum, The Hebrew University of Jerusalem, Israel, 2013.

* Contact author: a.ogawa@hsantalucia.it

 DOI:10.1163/22134808-000S0140

BRILL Multisensory Research 26 Supplement (2013) 188 brill.com/msr

Poster Presentation

Activity in superior colliculus during rubber hand illusion unravel network articulating body-part and full-body illusions processing: An fMRI study

Isadora Olivé *, **Claus Tempelmann and Hans-Jochen Heinze**
Universitatsklinikum Magdeburg A.o.R. Universitatsklinik fur Neurologie, Germany

Abstract
The experimental investigation of bodily self-consciousness assumes that body-part and full-body representations can be selectively assessed through body-part and full-body illusions as the rubber hand illusion (RHi) and the induced out-of-body experience (OBE), respectively. Nevertheless, we demonstrated in a previous work that induction of a body-part illusion is accompanied by the induction of a full-body illusion. In the current study we investigate the neuronal underpinnings of this link between body-part and full-body illusions. We used the traditional RHi paradigm in a functional magnetic resonance imaging (fMRI) study. Activity at the ipsilateral superior colliculus (SC) correlated with the illusion. This activation pattern is observed in hemispatial neglect, suggesting that the implication of superior colliculus in body-part illusion may underpin a correlated full-body illusion. To test this hypothesis, we applied a psychophysiological interaction (PPI) analysis aiming to unravel the connectivity of the superior colliculus during RHi. We observed a significant connectivity of ipsilateral superior colliculus with brain structures previously implicated in both body-part and full-body illusions, the bilateral primary somatosensory cortex, implicated with body-part illusions; the right temporoparietal junction (TPJ), associated with full-body illusions; the bilateral ventral premotor cortex (vPM), implicated both in body-part and full-body illusions; and the right fusiform body area (FBA), implicated in full-body representation. Altogether, our findings suggest that multisensory integration of bodily signals at the superior colliculus play a major role in emergence of bodily self-consciousness in humans, affecting both body-part and full-body representations.

Keywords
Bodily self-consciousness, body-part *vs.* full-body illusions, superior colliculus

Abstract from the 14th International Multisensory Research Forum, The Hebrew University of Jerusalem, Israel, 2013.
* Contact author: isadora.olive@med.ovgu.de

 DOI:10.1163/22134808-000S0141

BRILL Multisensory Research 26 Supplement (2013) 189 brill.com/msr

Poster Presentation

Cross-modal associations between vision, touch and audition influence visual search through top-down attention not bottom-up capture

Emily Orchard-Mills *, **Erik van der Burg and David Alais**
University of Sydney, Australia

Abstract
A recent study (Guzman-Martinez *et al.*, 2012) showed that participants match the frequency of an amplitude-modulated auditory stimulus to visual spatial frequency with a linear relationship and suggested this crossmodal mapping automatically guided attention to specific spatial frequencies. We replicated the reported matching relationship and also performed matching between tactile and visual spatial frequency. We then used the visual search paradigm to investigate whether auditory or tactile cues can guide attention to matched visual spatial frequencies. Participants were presented with a search display containing multiple Gabors, all with different spatial frequencies. When the auditory or tactile cue was informative, improved search efficiency occurred for some matched spatial frequencies, with the specificity of the effect being greater for touch than audition. However, when uninformative neither auditory and tactile cues produced any effect on visual search performance. Furthermore, when informative, unmatched auditory cues (shifted substantially from the reported match, but still matched in relative position) improved search performance. Taken together, these findings suggest that although auditory and tactile cues can influence visual selection of a matched spatial frequency, the effects are due to top-down attentional control rather than automatic attentional capture derived from low-level mapping.

Keywords
Visual search, attention, crossmodal, visual selection

Reference

Guzman-Martinez, E., Ortega, L., Grabowecky, M., Mossbridge, J. and Suzuki, S. (2012). Interactive coding of visual spatial frequency and auditory amplitude-modulation rate, *Current Biology: CB* **22**(5), 383–388. DOI:10.1016/j.cub.2012.01.004

Abstract from the 14th International Multisensory Research Forum, The Hebrew University of Jerusalem, Israel, 2013.
* Contact author: emilyorchardmills@gmail.com

DOI:10.1163/22134808-000S0142

BRILL Multisensory Research 26 Supplement (2013) 190 MULTISENSORY RESEARCH brill.com/msr

Poster Presentation

On pitch-elevation mapping. Nature, nurture and behaviour

Cesare V. Parise [1,2,*], **Katharina Knorre** [2] **and Marc O. Ernst** [2]

[1] Max Planck Institute for Biological Cybernetics, Germany

[2] Department of Cognitive Neuroscience, University of Bielefeld, Germany

Abstract

The association between auditory pitch and spatial elevation is one the most fascinating examples of cross-dimensional mappings: in a wide range of cognitive, perceptual, attentional and linguistic tasks, humans consistently display a positive, sometimes absolute, association between auditory pitch and spatial elevation. However, the origins of such a pervasive mapping are still largely unknown.

Through a combined analysis of environmental sounds and anthropometric measures, we demonstrate that, statistically speaking, this mapping is already present in both the distal and the proximal stimulus. Specifically, in the environment, high sounds are more likely to come from above; moreover, due to the filtering properties of the external ear, sounds coming from higher elevations have more energy at high frequencies.

Next, we investigated whether the internalized mapping depends on the statistics of the proximal, or of the distal stimulus. In a psychophysical task, participants had to localize narrow band-pass noises with different central frequencies, while head- and world-centred reference frames were put into conflict by tilting participants' body orientation. The frequency of the sounds systematically biased localization in both head- and world-centred coordinates, and, remarkably, in agreement with the mappings measured in both the distal and proximal stimulus.

These results clearly demonstrate that the cognitive mapping between pitch and elevation mirror the statistical properties of the auditory signals. We argue that, in a shorter time-scale, humans learn the statistical properties auditory signals; while, in a longer timescale, the evolution of the acoustic properties of the external ear itself is shaped by the statistics of the acoustic environment.

Keywords

Cross-dimensional mapping, statistics of the environment, auditory localization

Abstract from the 14th International Multisensory Research Forum, The Hebrew University of Jerusalem, Israel, 2013.

* Contact author: cesare.parise@tuebingen.mpg.de

DOI:10.1163/22134808-000S0143

BRILL Multisensory Research 26 Supplement (2013) 191–192 brill.com/msr

Poster Presentation

Moving my hand forward to touch my back

Polona Pozeg [1,2,*], **Giulio Rognini** [1,2] **and Olaf Blanke** [1,2]

[1] Laboratory of Cognitive Neuroscience, Brain Mind Institute, Ecole Polytechnique Fédérale de Lausanne, Lausanne, Switzerland

[2] Center for Neuroprosthetics, School of Life Science, Ecole Polytechnique Fédérale de Lausanne, Lausanne, Switzerland

Abstract

Bodily self-consciousness (BSC) can be altered by exposing subjects to ambiguous multisensory stimuli of one's body's appearance and location (Ehrsson, 2007; Lenggenhager *et al.*, 2007). In the somatic rubber hand illusion (sRHI) (Ehrsson *et al.*, 2005), a blindfolded participant passively strokes a rubber hand while the experimenter synchronously touches the participant's other hand. This tactile–proprioceptive conflict is resolved by integrating the simultaneous tactile stimuli into a unitary percept, resulting in the illusion of touching one's own hand (self-touch) and mislocalization of the stroked hand towards the rubber hand. Based on previous BSC work concerning full body representations (Lenggenhager *et al.*, 2007), we aimed to induce illusory self-touch at the trunk and observe changes in self-location.

Blindfolded participants ($n = 36$, 19 females) moved the master device of a robotic telemanipulator in front while the slave part applied tactile feedback to their back. In two experiments we manipulated the Delay between the participants' arm movements and tactile feedback, Resistance Force at the master device, and Movement Type (active/passive movement). Illusory self-touch was measured *via* questionnaires and self-location with a motor imagery task (Ionta *et al.*, 2011).

We found that illusory self-touch depended on the synchrony between the participant's arm movements and tactile feedback. Synchrony also resulted in forward self-location drift, suggesting a remapping of self-location towards the location of illusory self-touch (master device). Our data demonstrate that changes in BSC affecting the full body and its experienced position in space can be induced experimentally in the absence of visual input. These changes were induced despite the strong spatial incongruence between arm's sensorimotor signals and trunk somatosensory feedback.

Keywords

Sensorimotor integration, illusory self-touch, body representation

Abstract from the 14th International Multisensory Research Forum, The Hebrew University of Jerusalem, Israel, 2013.

* Contact author: polona.pozeg@epfl.ch

DOI:10.1163/22134808-000S0144

References

Ehrsson, H. (2007). The experimental induction of out-of-body experiences, *Science* **317**, 1048. DOI:10.1126/science.1142175

Ehrsson, H. H., Holmes, N. P. and Passingham, R. E. (2005). Touching a rubber hand: feeling of body ownership is associated with activity in multisensory brain areas, *J. Neurosci.* **25**, 10564–10573. DOI:10.1523/jneurosci.0800-05.2005

Ionta, S., *et al.* (2011). Multisensory mechanisms in temporo-parietal cortex support self-location and first-person perspective, *Neuron* **70**, 363–374.

Lenggenhager, B., Tadi, T., Metzinger, T. and Blanke, O. (2007). Video ergo sum: manipulating bodily self-consciousness, *Science* **317**, 1096–1099. DOI:317/5841/1096 [pii] 10.1126/science.1143439

BRILL Multisensory Research 26 Supplement (2013) 193 brill.com/msr

Poster Presentation

Stable body-centered coding of touch on the back

Lisa M. Pritchett * **and Laurence R. Harris**

Centre for Vision Research, York University, Canada

Abstract

Head position affects touch localization on the torso in different directions depending whether the head returns to centre or remains eccentric during testing. These observations have been taken to suggest that touch is coded in either gaze-centered or body-centered reference frames depending on the task (Pritchett *et al.*, 2012). Here we tested localization of tactile stimuli applied to the back using the same paradigms to indicate gaze-centered or body-centered coding. Tactile stimulation was applied using a horizontal array of eight tactors mounted on a belt, centred on the backbone. Participants reported the perceived location of touches on a visual scale presented on a screen. The head was held eccentrically during the stimulation and either returned to centre or remained eccentric during response. When the head was returned to centre, as for the touches on the front of the torso, perceived touch locations on the back were shifted in the head's direction, with largest effects on the side to which the head was turned. When the head remained eccentric, perceived touch location was unaffected by head position. Returning the head to centre evokes a gaze-centred strategy with errors related to errors in coding gaze. Leaving the head eccentric is associated with a body-centred coding strategy. The lack of errors on the back, compared to on the front of the torso under this condition may indicate a body-centred representation of the back that is more stable than the front — perhaps related to the availability of the backbone as a prominent body landmark.

Keywords

Touch localization, reference frames, body representation

Reference

Pritchett, L. M., Carnevale, M. J. and Harris, L. R. (2012). Reference frames for coding touch location depend on the task, *Experimental Brain Research* **222**, 437–445.

Abstract from the 14th International Multisensory Research Forum, The Hebrew University of Jerusalem, Israel, 2013.

* Contact author: lmpritch@yorku.ca

 DOI:10.1163/22134808-000S0145

BRILL Multisensory Research 26 Supplement (2013) 194 brill.com/msr

Poster Presentation

Audiovisual temporal integration in adults with Autism Spectrum Disorder

Paula Regener *
University of Glasgow, UK

Abstract
The ability to integrate auditory and visual information is important for everyday life. The Temporal Integration Window (TIW) measures how much asynchrony can be tolerated between auditory and visual streams before one loses the perception of a unitary audiovisual event. Previous investigations of TIW in individuals with Autism Spectrum Disorders (ASD) show mixed results in how performance compares to typically developed individuals (TD). The current study looked at TIW across a range of audiovisual stimuli to further examine this issue. The stimuli included the following audiovisual pairings: (1) a beep with a flashing circle (BF), (2) a point-light drummer with a drumbeat (PLD), (3) a face moving to say a single word and the voice saying the word (FV).

Eleven adult males with ASD, and their age, sex and IQ matches were shown the three audiovisual stimuli with varying degrees of audiovisual asynchrony. In separate blocks participants were asked to make either Temporal Order Judgements (TOJ) or Synchrony Judgements (SJ) when presented with these stimuli. For both TOJ and SJ psychophysical fits to the data provided estimates of the Point of Subjective Synchrony (PSS) and the width of TIW.

Individual ANOVAs on TIW and PSS for every stimulus type were ran using a within factor of judgement (SJ, TOJ) and a between factor of group (ASD, TD). These revealed no group differences in TIW or PSS. However, the FV TIW was significantly wider for TOJ than SJ. Additionally, PSS for all stimuli was significantly influenced by the types of judgements.

Keywords
Audiovisual temporal integration, autism spectrum disorder

Abstract from the 14th International Multisensory Research Forum, The Hebrew University of Jerusalem, Israel, 2013.

* Contact author: p.regener.1@research.gla.ac.uk

DOI:10.1163/22134808-000S0146

BRILL Multisensory Research 26 Supplement (2013) 195–196 brill.com/msr

Poster Presentation

'Visual'-parsing without visual experience: Is 'seeing' with sounds better than sight restoration in the early blind?

Lior Reich [1,*], **Miriam Guendelman** [1] **and Amir Amedi** [1,2]

[1] Department of Medical Neurobiology, The Institute for Medical Research Israel–Canada, Faculty of Medicine, The Hebrew University of Jerusalem, Israel
[2] The Edmond and Lily Safra Center for Brain Sciences (ELSC), The Hebrew University of Jerusalem, Israel

Abstract

Early-blind individuals who regained sight medically in adulthood, demonstrated that functional-vision acquisition in adulthood is challenging. For instance, Ostrovsky *et al.* (2009) reported poor behavior in static-visual-parsing (i.e., segregating the image into distinct unified objects), a critical visual concept and an ability required for performing any visual task in natural-environments.

Another rehabilitation approach is to use non-invasive Sensory-Substitution-Devices (SSDs), which transform visual images into sounds. Various abilities were demonstrated using SSDs, however 'visual'-parsing was never directly tested.

Here we tested whether congenitally-blind adults can learn to perform 'visual'-parsing with SSD, using similar stimuli and approach to those used by Ostrovsky *et al.* (2009), but this time with the visual information conveyed through sounds; and compared the SSD users' performance to that reported for the medically-sight-restored individuals.

The SSD users performed significantly above chance-level, following only ~70 training hours. Interestingly, they outperformed the sight-restored-individuals, who had months of constant eyesight, in all tasks tested.

In a second test, we found that the SSD users could discern the vantage-point of 3D objects from 2D SSD images, a task requiring correct parsing (at least in the local-level). Theoretically, the results demonstrate that the adult brain retains visual learning capacity; and suggests that with adequate training and technologies some high-order visual aspects can be acquired in adulthood, even without any visual-experience during developmental critical-periods. Practically, the results support the potential use of SSDs as standalone daily-aids, but also suggest a potential for combining invasive-restoration approaches with SSD input and/or training to improve and enhance rehabilitation.

Abstract from the 14th International Multisensory Research Forum, The Hebrew University of Jerusalem, Israel, 2013.
* Contact author: lioreich@gmail.com

 DOI:10.1163/22134808-000S0147

Keywords
Sight restoration, early blind

Reference

Ostrovsky, Y., Meyers, E., Ganesh, S., Mathur, U. and Sinha, P. (2009). Visual parsing after recovery from blindness, *Psychological Science* **20**(12), 1484–1491.

BRILL Multisensory Research 26 Supplement (2013) 197–198 brill.com/msr

Poster Presentation

The effect of ageing on acoustic facilitation of object movement detection within optic-flow

Eugenie Roudaia [1,*], **Finnegan J. Calabro** [2], **Lucia M. Vaina** [2] **and Fiona N. Newell** [1]

[1] Trinity College Dublin, Ireland
[2] Boston University, Boston, MA, USA

Abstract

Multisensory integration appears to be enhanced in older age (e.g., Laurienti *et al.*, 2006; Maguinness *et al.*, 2011). Given that motion perception declines with ageing, we examined whether multisensory integration may enhance motion perception in older adults. Calabro *et al.* (2011) recently showed that the detection of object motion in the context of self-motion is facilitated by the presentation of a co-localized and directionally congruent auditory stimulus. We compared performance of younger (mean age: 26 years) and older (mean age: 68 years) participants on this task. On each trial, nine identical spheres were randomly distributed within 10 degrees of fixation and 8 spheres expanded outward for 1 s, simulating forward translation at a constant speed. One sphere (the target) had independent motion either expanding or receding at one of three different speeds. A response screen prompted the participants to report which of four highlighted spheres was the target. The task was presented with no sound and with a co-localized sound that simulated motion congruent with the target's direction. Older participants showed overall poorer target detection than younger participants. Consistent with Calabro *et al.* (2011), the auditory stimulus improved target detection in younger participants, however, it had less of an effect on older participants' performance. A control experiment confirmed that both groups were able to localize the auditory stimulus, however localization was poorer in the older group. Future studies equating unisensory performance in younger and older participants are needed to determine whether multisensory integration in motion is affected in older age.

Keywords

Motion, ageing, audiovisual motion, multisensory enhancement

Abstract from the 14th International Multisensory Research Forum, The Hebrew University of Jerusalem, Israel, 2013.

* Contact author: roudaiae@tcd.ie

 DOI:10.1163/22134808-000S0148

References

Calabro, F. J., Soto-Faraco, S. and Vaina, L. M. (2011). Acoustic facilitation of object movement detection during self-motion, *Proc. Biol. Sci.* **278**(1719), 2840–2847.

Laurienti, P. J., Burdette, J. H., Maldjian, J. A. and Wallace, M. T. (2006). Enhanced multisensory integration in older adults, *Neurobiology of Aging* **27**(8), 1155–1163.

Maguinness, C., Setti, A., Burke, K. E., Kenny, R. A. and Newell, F. N. (2011). The effect of combined sensory and semantic components on audio-visual speech perception in older adults, *Front Aging Neurosci.* **3**, 19.

BRILL Multisensory Research 26 Supplement (2013) 199–200 brill.com/msr

Poster Presentation

Visuo-vestibular interaction in the perception of curvilinear trajectories for passive and active drivers

Florian Savona [1,*], **Anca M. Stratulat** [2], **Christophe Bourdin** [1] **and Vincent Roussarie** [2]

[1] Aix Marseille University, CNRS, ISM, France
[2] PSA Peugeot–Citroën, France

Abstract

Self-motion perception, which partly determines the realism of dynamic driving simulators, is based on multisensory integration. For curved trajectories, adding linear translations to visual stimulation seems to improve the perception of motion (Bertin *et al.*, 2004). However, cornering means not only translational motion but also rotational motion in yaw. Wikie and Wand (2005) found no effect of yaw motion on steering performance, but it seems that higher rates of physical stimulation could have a more visible effect. However, the influence of yaw acceleration on cornering perception is still a matter of debate, especially when associated to optic flow. Therefore, the present study aims to analyze the respective role of vestibular (yaw acceleration) and visual stimulations for the perception of curvilinear trajectories. We designed two experiments in which the subjects had 1/to orally estimate their angular displacements as passive drivers, 2/to generate angular displacements by controlling the steering wheel as active drivers. In both experiments, subjects were submitted to three different conditions: (1) visual motion, (2) physical yaw acceleration, (3) combined visual and physical motions. Preliminary results of the first experiment show that visual stimulation produces greater overestimations of angular displacements than physical yaw motion, this overestimation being in between in the visuo-vestibular condition. They also suggest that the weights of visual and vestibular cues in cornering perception depend on the amplitude of the angular displacements. The second experiment should allow us to observe the evolution of visuo-vestibular interaction when subjects are active rather than passive drivers.

Keywords

Visuo-vestibular interaction, curvilinear trajectories, yaw motion

Abstract from the 14th International Multisensory Research Forum, The Hebrew University of Jerusalem, Israel, 2013.

* Contact author: savonaflorian@gmail.com

 DOI:10.1163/22134808-000S0149

References

Bertin, R. J. V. and Berthoz, A. (2004). Visuo-vestibular interaction in the reconstruction of travelled trajectories, *Experimental Brain Research* **154**(1) (Janvier 1, 2004), 11–21. DOI:10.1007/s00221-003-1524-3

Wilkie, R. M. and Wann, J. P. (2005). The role of visual and nonvisual information in the control of locomotion, *Journal of Experimental Psychology. Human Perception and Performance* **31**(5) (Octobre 2005), 901–911. DOI:10.1037/0096-1523.31.5.901

BRILL Multisensory Research 26 Supplement (2013) 201–202 brill.com/msr

Poster Presentation

External spatial coordinates for tactile spatial attention are reflected in prestimulus oscillations in sighted but not in blind individuals

Jonathan T. W. Schubert [1,*], Tobias Heed [1], Julia Föcker [2] and Brigitte Röder [1]

[1] Biological Psychology and Neuropsychology, University of Hamburg, Germany
[2] Department of Psychology and Educational Sciences, University of Geneva, Switzerland

Abstract

Sighted individuals seem to represent skin locations automatically in anatomical and external reference frames (Shore *et al.*, 2002; Yamamoto and Kitazawa, 2001), which are reflected in lateralized oscillatory activity in the beta and alpha range, respectively (Buchholz *et al.*, 2013). Congenitally blind individuals do not seem to automatically activate external reference frames suggesting a crucial contribution of developmental vision for touch localization.

We recorded EEG while sighted and congenitally blind participants directed attention to one hand in order to detect rare tactile targets at this hand only. In each trial, the relevant hand was indicated by an auditory cue (S1) which preceded the tactile stimulus (S2) by 1000 ms. Participants either adapted an uncrossed or a crossed hand posture. Attention- and posture-related power modulations in the alpha band were analyzed.

In the sighted orienting attention to one uncrossed hand (S1–S2 interval) was accompanied by an alpha reduction over contralateral parietal cortex and an alpha increase over ipsilateral cortex. With hands crossed a contralateral alpha decrease (relative to the anatomical hand) was still observed, but the ipsilateral alpha increase was less pronounced.

In the blind orienting attention to one of the hands was accompanied by a contralateral alpha decrease over parietal cortex which extended towards the ipsilateral hemisphere as well. By contrast, an alpha increase was not observed. This pattern of alpha activation in the congenital blind remained unchanged when they adopted a crossed hand posture.

These findings complement ERP evidence about deployment of attention in sighted and blind (Eardley and van Velzen, 2011) and extend previous evidence that alpha activity related mechanisms of attention control depend on developmental vision.

Abstract from the 14th International Multisensory Research Forum, The Hebrew University of Jerusalem, Israel, 2013.

* Contact author: jonathan.schubert@uni-hamburg.de

 DOI:10.1163/22134808-000S0150

References

Buchholz, V. N., Jensen, O. and Medendorp, W. P. (2013). Parietal oscillations code nonvisual reach targets relative to gaze and body, *The Journal of Neuroscience* **33**(8), 3492–3499.

Eardley, A. F. and van Velzen, J. (2011). Event-related potential evidence for the use of external coordinates in the preparation of tactile attention by the early blind, *The European Journal of Neuroscience* **33**(10), 1897–1907.

Shore, D. I., Spry, E. and Spence, C. (2002). Confusing the mind by crossing the hands, *Brain Research, Cognitive Brain Research* **14**(1), 153–163.

Yamamoto, S. and Kitazawa, S. (2001). Reversal of subjective temporal order due to arm crossing, *Nature Neuroscience* **4**(7), 759–765.

BRILL Multisensory Research 26 Supplement (2013) 203 brill.com/msr

Poster Presentation

The marble-hand illusion

Irene Senna [1,*], Angelo Maravita [1], Nadia Bolognini [1] and Cesare V. Parise [2,3]

[1] Milano-Bicocca University, Italy

[2] Max Planck Institute for Biological Cybernetics, Germany

[3] Department of Cognitive Neuroscience, University of Bielefeld, Germany

---

Abstract

Our body is made of flesh and bones. We know it, and in our daily lives all the senses — including touch, vision, and audition — constantly provide converging information about this simple, factual truth. But is this necessarily always the case? Here we report a surprising bodily illusion demonstrating that human observers rapidly update their assumptions about the material qualities of their body, based on their recent multisensory perceptual experience. To induce an illusory misperception of the material properties of the hand, we repeatedly gently hit participants' hand, while progressively replacing the natural sound of the hammer against the skin with the sound of a hammer hitting a piece of marble. After five minutes, the hand started feeling stiffer, heavier, harder, less sensitive, and unnatural, and showed enhanced Galvanic skin response to threatening stimuli. This novel bodily illusion, the 'Marble-Hand Illusion', demonstrates that the experience of the material of our body, surely the most stable attribute of our bodily self, can be quickly updated through multisensory integration.

Abstract from the 14th International Multisensory Research Forum, The Hebrew University of Jerusalem, Israel, 2013.

* Contact author: i.senna@campus.unimib.it

 DOI:10.1163/22134808-000S0151

BRILL Multisensory Research 26 Supplement (2013) 204 brill.com/msr

Poster Presentation

The development of cross-modal processing in the primary somatosensory cortex of the rat

Kay Sieben [1,*], Brigitte Röder [2] and Ileana L. Hanganu-Opatz [1]

[1] University Medical Center Hamburg-Eppendorf, Center for Molecular Neurobiologie Hamburg, Germany

[2] Biological Psychology and Neuropsychology, University Hamburg, Germany

Abstract

Multisensory interactions have recently been demonstrated in putatively unisensory neocortical regions including the primary cortices. While the aquisition of multisensory functions and maturation of cross-modal processing in high order convergence regions (e.g., superior colliculus) critically depend on minimal multisensory experience during defined periods, the developmental mechanisms of cross-modal integration in primary sensory cortices are still unknown. We focused on the convergence of visual and tactile information in the primary somatosensory (S1) and visual (V1) cortices of Brown Norway rat. Light flashes and whisker deflections were combined with multi-site extracellular recordings from S1 and V1 *in vivo*. First, we assessed whether minimal cross-modal exploration is mandatory for the maturation of multisensory functions by focussing on the pre-juvenile developmental stage (retina light sensitive, eyes closed, no whisking). Similar to previous data from adult rats, cross-modal processing in the S1 of pre-juvenile rats relies on modulation of network oscillations *via* cortico-cortical projections. Second, we elucidated the role of unisensory experience for cross-modal processing by cutting the whiskers of neonatal rats and thus, depriving them of tactile inputs during the critical period. The sensory deprivation significantly changed the cross-modal processing in the S1. Consequently, we propose unisensory development rather than active cross-modal exploration is mandatory for the correct development of multisensory processing in primary cortical areas.

Abstract from the 14th International Multisensory Research Forum, The Hebrew University of Jerusalem, Israel, 2013.

* Contact author: kay.sieben@zmnh.uni-hamburg.de

DOI:10.1163/22134808-000S0152

BRILL Multisensory Research 26 Supplement (2013) 205 brill.com/msr

Poster Presentation

The neurochronometry of reading in the blind

Nadine Sigalov [1,*], **Laurent Cohen** [2,3,4,5] **and Amir Amedi** [1,6]

[1] Department of Medical Neurobiology, Institute for Medical Research Israel–Canada (IMRIC), The Hebrew University, Jerusalem, Israel

[2] AP-HP, Groupe hospitalier Pitié-Salpêtrière, Department of Neurology, Paris, France

[3] INSERM, ICM Research Center, UMRS 975, Paris, France

[4] University Paris 6, Faculté de Médecine Pitié-Salpêtrière, Paris, France

[5] Institut du Cerveau et de la Moëlle épinière, Paris, France

[6] The Edmond and Lily Safra Center for Brain Sciences (ELSC), The Hebrew University, Jerusalem, Israel

Abstract

Using a visual-to-auditory sensory-substitution algorithm, congenitally fully blind adults were taught to read letters and words using 'soundscapes' — sounds topographically representing images. High resolution fMRI was used to examine key questions regarding the neurochronometry of the reading process in the blind: the areas which are involved in the reading network of the blind in the different phases of the process and the onset latencies of the different areas participating in the network. The central foci of the reading network of the blind were found to be a subset of the areas involved in the reading of degraded stimuli. Further, the onset of the reading process begins in the auditory cortex and then moves to occipital cortex and roughly follows the processing hierarchy expected in the sighted individuals.

Keyword

Neurochronometry

Abstract from the 14th International Multisensory Research Forum, The Hebrew University of Jerusalem, Israel, 2013.

* Contact author: nadine_sigalov@yahoo.com

 DOI:10.1163/22134808-000S0153

BRILL Multisensory Research 26 Supplement (2013) 206 brill.com/msr

Poster Presentation

Switching between the senses: Imagination interferes with perception in a stimulus detection tasks

Benjamin Steinweg [1,*], Fred Mast [1] and Joseph Krummenacher [2]

[1] University of Bern, Switzerland

[2] Ludwig-Maximilians-Univesität München, Germany

Abstract

In an attempt to examine whether stimulus detection in a simple reaction time task is expedited by an imagined stimulus, participants were asked to imagine a familiar stimulus in one sensory modality (e.g., sound) while waiting to signal the onset of a target stimulus in the other modality (e.g., vision). Contrary to what could be expected from research into multi-sensory integration and what was replicated in our baseline conditions where physical signals were presented, imagination of an additional stimulus did not lead to redundancy gains, but decreased response speed. The result is explicable by a mechanism that requires the dynamic allocation of a limited processing capacity — attentional weight — to sensory modalities, which results in reaction time costs when observers are required to switch from imagination in one modality to perception in the other modality. In two follow-up experiments the imagination task was exactly the same as in experiment 1, but the signal to be detected was either a bimodal signal or a signal of the same modality as the signal to be imagined. Here, no RT costs were found compared to the respective baseline conditions where the target to be detected was the same but no imagination task was to be executed. Consequently the effect found in experiment 1 does not appear to be due to an increase in unspecific cognitive load, task switching or task coordination demands but due to the switch in modalities between the imagery induced working memory content and the physically perceived detection target.

Keywords

Modality switching, imagery, attention, integration

Abstract from the 14th International Multisensory Research Forum, The Hebrew University of Jerusalem, Israel, 2013.

* Contact author: benjamin.steinweg@psy.unibe.ch

 DOI:10.1163/22134808-000S0154

BRILL Multisensory Research 26 Supplement (2013) 207 brill.com/msr

Poster Presentation

Deficient multisensory integration in schizophrenia: An event-related potential study

Jeroen Stekelenburg *

Tilburg University, The Netherlands

Abstract

The N1 component of the event-related brain potential (ERP), reflecting neural activity associated with basic sound processing, can be suppressed if a sound is accompanied by a video that predicts sound onset. In this study we examined whether visual predictive information induces auditory de-activation in patients with schizophrenia. The electroencephalography (EEG) was recorded in 18 patients with schizophrenia and compared with that of 18 healthy volunteers. As stimuli we used video recordings of natural actions in which visual information preceded and predicted the onset of the concordant sound (e.g., a video recording of hand clapping). For the healthy control group, visual information substantially reduced the auditory-evoked N1 if compared to a sound-only condition. This reduction of the N1 was absent in patients with schizophrenia. The results show a deficit in multisensory processing in patients with schizophrenia and suggest that multisensory integration dysfunction may be an important and, to date, under-researched aspect of schizophrenia.

Keywords

Schizophrenia, event-related brain potentials, auditory N1, multisensory perception, audiovisual integration, temporal prediction

Abstract from the 14th International Multisensory Research Forum, The Hebrew University of Jerusalem, Israel, 2013.

* Contact author: J.J.Stekelenburg@uvt.nl

DOI:10.1163/22134808-000S0155

BRILL

Multisensory Research 26 Supplement (2013) 208

MULTISENSORY RESEARCH

brill.com/msr

Poster Presentation

fMRI-guided TMS of the superior temporal sulcus impairs multisensory temporal processing

Ryan A. Stevenson [1,*], **Caitlin R. Mullin** [2,3], **Mark T. Wallace** [1] **and Jennifer K. E. Steeves** [2]

[1] Vanderbilt Universtiy Medical Center, Nashville, TN, USA

[2] York University, Canada

[3] University of Leuven, Canada

Abstract

The temporal coincidence of multisensory inputs plays a strong role in perceptual binding; the more temporally proximate two stimuli are, the higher the likelihood of binding. The superior temporal sulcus (STS) has been implicated in binding audiovisual stimuli and temporal processing. Here, we used fMRI-guided repetitive transcranial magnetic stimulation (rTMS) to assess the contribution of STS to multisensory temporal processing. A pre-TMS anatomical and functional MRI was acquired, and multisensory (A ∩ V, STS), auditory (A > V, Hechl's gyrus), and visual (V > A, striate cortex) regions of interest (ROIs) were localized. During separate sessions, each of these ROIs was stimulated for 20 min at 1 Hz prior to collection of behavioral data, in addition to a no-TMS condition. For the behavioral task, participants were presented with flash-beep stimuli at parametrically-varied stimulus-onset asynchronies (SOAs) while performing a simultaneity judgment task. Temporal binding windows (TBW; the range of SOAs within which a participant perceived the flash and beep to be synchronous 50% of the time) were calculated from these responses. Relative to the no-TMS condition, stimulation of auditory cortex induced a broadening of the TBW with auditory-leading stimuli, and stimulation of visual cortex induced a broadening of the TBW with visual-leading stimuli. Stimulation of STS induced a significant change in the full TBW, driven primarily by an increased tolerance for visual-leading stimuli. Changes specific to the more ecologically valid visual-leading stimuli with STS disruption suggest that multisensory temporal processing in STS reflects learned environmental statistics.

Keywords

Temporal processing, TMS, fMRI, superior temporal sulcus, synchrony, psychophysics, multisensory, audiovisual

---

Abstract from the 14th International Multisensory Research Forum, The Hebrew University of Jerusalem, Israel, 2013.

* Contact author: ryan.andrew.stevenson@gmail.com

DOI:10.1163/22134808-000S0156

BRILL Multisensory Research 26 Supplement (2013) 209 brill.com/msr

Poster Presentation

Interactions of multisensory integration and spatial attention in auditory space perception

Marc Stöhr *, **Jan Bennemann, Claudia Freigang and Rudolf Rübsamen**

University of Leipzig, Germany

Abstract

In reverberant environments sound reflections (echoes) interfere with spatial information of the sound source. The precedence effect is a mechanism that suppresses spatial information of echoes arriving with a short latency and fuses them with the sound source, helping listeners to localize sound sources. Two recent papers showed that the strength of the precedence effect can be modulated by spatially and temporally congruent visual information (Bishop *et al.*, 2011) and exogenous spatial attention (London *et al.*, 2012). If a concurrent visual stimulus or a visual cue (300 ms before auditory stimulus) occurred at the location of the sound source, echo suppression was enhanced. If the respective visual stimuli occurred at the location of the echo, the suppression mechanism was inhibited. While multisensory integration was earlier thought to be an automatic process not influenced by attention, in recent years several studies were able to show that attention can influence multisensory integration under specific conditions. In this study we combine visual (multisensory context) and exogenous attentional modulation of echo suppression to examine how interactions of multisensory integration and bottom-up driven attentional modulation shape auditory space perception. From the comparison of listener's performance in trials with either (i) visual or (ii) attentional modulation of echo suppression alone or with (iii) simultaneous visual and attentional modulation to either the same or different locations the interrelation of the two processes can be disentangled.

Keywords

Multisensory integration, spatial attention, echo suppression, auditory system

References

Bishop, C. W., London, S. and Miller, L. M. (2011). Visual influence on echo suppression, *Curr. Biol.* **21**, 221–225.

London, S., Bishop, C. W. and Miller, L. M. (2012). Spatial attention modulates the precedence effect, *J. Exp. Psychol. Hum. Percept. Perform.* **38**, 1371–1379.

Abstract from the 14th International Multisensory Research Forum, The Hebrew University of Jerusalem, Israel, 2013.

* Contact author: marc.stoehr@uni-leipzig.de

 DOI:10.1163/22134808-000S0157

BRILL Multisensory Research 26 Supplement (2013) 210 brill.com/msr

Poster Presentation

An extrastriate body-selective area in the congenitally blind

Ella Striem-Amit [1,*] **and Amir Amedi** [1,2]

[1] Department of Medical Neurobiology, The Institute for Medical Research Israel–Canada, Faculty of Medicine, The Hebrew University of Jerusalem, Israel

[2] The Edmond and Lily Safra Center for Brain Sciences (ELSC), The Cognitive Science Program, The Hebrew University of Jerusalem, Jerusalem, Israel

Abstract

Vision is by far the most prevalent modality for experiencing others' body postures and motion, and thus their actions and intents. The absence of visual experience therefore poses a unique situation for the creation of the brain's representation of body information. Would a congenital absence of this information, generating also a lack of expertise and exposure to body shapes (except from very limited experience *via* touch) still allow for the development of the body-selective areas? And where may body-selectivity be found? Would it be limited to the parieto-frontal cortices, based on the input modality for such information in the blind? We investigated this question using a visual-to-auditory sensory-substitution algorithm, enabling congenitally fully-blind adults to perceive full-body configurations. We find that despite the sole exposure of body shape information *via* the somatosensory modality, the blind activated primarily the visual cortex, and specifically the location of the Extrastriate Body Area (EBA) during the perception of body shapes. Moreover, the blind showed selectivity of the right inferior temporal sulcus, in the general anatomical location of the extrastriate body area, for the perception of body-shapes over textures, objects and even faces as is seen in normally sighted people. Thus, the EBA region may develop and engage in its role in perception of body-information in the absence of vast exposure to body-shape information, and even in the absence of visual experience.

Keywords

Blindness, body representation, cross-modal, development, plasticity, vision

Abstract from the 14th International Multisensory Research Forum, The Hebrew University of Jerusalem, Israel, 2013.

* Contact author: ella.striem@mail.huji.ac.il

DOI:10.1163/22134808-000S0158

BRILL Multisensory Research 26 Supplement (2013) 211 brill.com/msr

Poster Presentation

Bouncing human *vs.* bouncing balls: Synchrony perception between auditory rhythm and different naturalistic visual motions

Yi-Huang Su *

Department of Movement Science, Technical University Munich, Germany

Abstract

Temporal integration of auditory and visual information is often examined in scenarios where sounds are produced by the observed action (e.g., impact sounds paired with drumming movements). The present study, however, investigated synchrony perception between concurrent auditory and visual streams when the sounds are not consequent upon the movement (e.g., watching a dancer moving to the music), and how visual motion velocity affects this perception. In Experiment 1, participants judged synchrony at the bounce between a bouncing point-light figure (PLF) and a regular auditory rhythm in four visual conditions: human PLF bouncing with human motion, human PLF bouncing with ball motion (without damping), non-human PLF bouncing with human motion, and non-human PLF bouncing with ball motion. Regardless of the figure, a greater delay of the auditory stream was found for synchrony perception when viewing movements of ball motion than of human motion. This suggests a shift of point of subjective synchrony (PSS) towards the maximal velocity change within the observed trajectory, which occurs later in ball motion than in human motion. Experiment 2 measured synchrony perception either at the bounce or at the highest position of the same movement. The effect of motion on PSS was found again for synchrony judgement at the bounce, but not at the highest position, where velocity change was similar in both motions. Together these results demonstrate the effect of visual velocity cue — specifically the point of peak velocity change in the trajectory — on audiovisual synchrony perception in a continuous naturalistic movement.

Keywords

Audiovisual rhythm, biological motion, synchrony perception

Abstract from the 14th International Multisensory Research Forum, The Hebrew University of Jerusalem, Israel, 2013.

* Contact author: yihuang.su@tum.de

DOI:10.1163/22134808-000S0159

BRILL Multisensory Research 26 Supplement (2013) 212–213 brill.com/msr

Poster Presentation

Concurrent sensorimotor adaptation for different delays measured *via* a tapping task

Yoshimori Sugano [1,*], **Mirjam Keetels** [2] **and Jean Vroomen** [2]

[1] Kyushu Sangyo University, Japan
[2] Tilburg University, The Netherlands

Abstract

Perception of temporal synchrony between one's own action and the sensory feedback of that action is quite flexible (Heron *et al.*, 2009; Keetels and Vroomen, 2012; Stekelenburg *et al.*, 2011; Stetson *et al.*, 2006; Sugano *et al.*, 2010, 2012). We examined whether concurrent sensory-motor temporal recalibration to different lags is possible for the left and right hand. During exposure, participants tapped their left and right index fingers in alternating fashion while each tap induced an auditory feedback signal (a short click sound). One hand was exposed to a long delay between the tap and the sound (∼150 ms), while the other hand was exposed to a subjective no-delay (∼50 ms) ('adaptation'). After the adaptation, participants then performed a synchronous tapping task ('post-test') in which they tried to tap in synchrony with pacer tones presented at a constant inter-stimulus interval (ISI = 1000 ms). The results showed that the hand that was exposed to the delayed sound corrected for this delay by tapping earlier (a larger anticipation error) than the no-delayed hand. This temporal recalibration of the delayed hand occurred faster and was more complete (∼40%) if the two hands were exposed to the same rather than different delays (∼20%). This indicates that there is a central and a motor-specific component in motor-sensory temporal recalibration. Tapping in synchrony with a pacing signal is thus a sensitive measure of temporal recalibration with central and motor-specific components.

Keywords

Sensory-motor adaptation, crossmodal processing, time perception

References

Heron, J., Hanson, J. V. and Whitaker, D. (2009). Effect before cause: Supramodal recalibration of sensorimotor timing, *PLoS One* **4**, e7681.

Keetels, M. and Vroomen, J. (2012). Exposure to delayed visual feedback of the hand changes motor-sensory synchrony perception, *Experimental Brain Research* **219**(4), 431–440.

Abstract from the 14th International Multisensory Research Forum, The Hebrew University of Jerusalem, Israel, 2013.

* Contact author: robo1972@gmail.com

 DOI:10.1163/22134808-000S0160

Stekelenburg, J. J., Sugano, Y. and Vroomen, J. (2011). Neural correlates of motor-sensory temporal recalibration, *Brain Research* **1397**, 46–54.

Stetson, C., Cui, X., Montague, P. R. and Eagleman, D. M. (2006). Motor-sensory recalibration leads to an illusory reversal of action and sensation, *Neuron* **51**, 651–659.

Sugano, Y., Keetels, M. and Vroomen, J. (2010). Adaptation to motor-visual and motor-auditory temporal lags transfer across modalities, *Experimental Brain Research* **201**, 393–399.

Sugano, Y., Keetels, M. and Vroomen, J. (2012). The build-up and transfer of sensorimotor temporal recalibration measured *via* a synchronization task, *Frontiers in Psychology* **3**(246). DOI:10.3389/fpsyg.2012.00246

BRILL Multisensory Research 26 Supplement (2013) 214 brill.com/msr

Poster Presentation

The influence of temporal regularities and cross-modal temporal cues on auditory detection

Sanne ten Oever [1,*], Charles Schroeder [2,3], David Poeppel [4], Nienke van Atteveldt [5] and Elana Zion-Golumbic [3]

[1] Department of Cognitive Neuroscience, University Maastricht, The Netherlands
[2] The Nathan Kline Institute, Orangeburg, NY, USA
[3] Columbia University, New York, NY, USA
[4] New York University, New York, NY, USA
[5] Netherlands Institute for Neuroscience & Maastricht University, The Netherlands

Abstract

Temporal structure in the environment often has predictive value for anticipating the occurrence of forthcoming events. In this study we compared the influence of two types of predictive temporal information on auditory perception: (1) intrinsic temporal rhythmicity of an auditory stimulus stream and (2) temporally-predictive visual cues. We hypothesized that combining predictive temporal information within- and across-modality would improve auditory detection, beyond the advantage provided by each information source alone. We presented streams of tones at either increasing or decreasing intensities until participants reported that they could hear/no longer hear the tones. Tones were presented in either rhythmic or random sequences and were preceded by a temporally predictive visual flash in half of the trials. We show that detection thresholds are lower for rhythmic (*vs.* random) and audiovisual (*vs.* auditory-only) presentation, and that this effect is additive for rhythmic audiovisual presentation in both paradigms. These behavioral results suggest that both types of temporal information are used in parallel to prepare the perceptual system for upcoming stimuli and to optimally interact with the environment. Our findings underscore the flexibility and proactivity of the perceptual system which uses these temporal contextual factors combined to anticipate upcoming events and process them optimally.

Keywords

Temporal prediction, audiovisual integration, rhythmicity, detection

---

Abstract from the 14th International Multisensory Research Forum, The Hebrew University of Jerusalem, Israel, 2013.

* Contact author: sanne.tenoever@maastrichtuniversity.nl

 DOI:10.1163/22134808-000S0161

BRILL Multisensory Research 26 Supplement (2013) 215 brill.com/msr

Poster Presentation

Distortion of auditory space during visually induced self-acceleration

Wataru Teramoto [1,*], Zhenglie Cui [2], Kazuki Moishi [1], Shuichi Sakamoto [2], Yo-iti Suzuki [2] and Jiro Gyoba [3]

[1] Muroran Institute of Technology, Japan
[2] Research Institute of Electrical Communication, Tohoku University, Japan
[3] Department of Psychology, Tohoku University, Japan

Abstract

Perception of self-motion is based on integration of multiple sensory inputs, especially from the vestibular and visual systems. Our previous study demonstrated that vestibular information on linear acceleration distorted auditory space perception (Teramoto *et al.*, 2012, *PLoS ONE*, **7**(6), e39402). However, it is unclear whether this phenomenon is contingent on vestibular signals or can be caused by inputs from the other sensory modalities involving self-motion perception. Here, we investigated whether visual information on linear self-acceleration can also alter auditory space perception. Large-field visual motion which simulated either forward or backward self-motion was presented on a screen without vestibular information. During the visual motion stimulation, a short noise burst was delivered from one of loudspeakers that were aligned parallel to the simulated self-motion direction along a wall to the left of participants. The participants indicated in which direction the noise burst was presented, forward or backward relative to their subjective coronal plane. The noise burst was delivered after participants reported self-motion (Experiment 1) or irrespective of the participants' experience of self-motion (Experiment 2). Results showed that the sound position aligned with the subjective coronal plane was displaced in the traveling direction for self-acceleration conditions only when the sound was delivered after the participants reported self-motion perception. These results suggest that self-motion information, irrespective of its origin, is crucial for this distortion of auditory space perception.

Keywords

Self-motion, auditory space perception, vection

Reference

Teramoto, W., Sakamoto, S., Furune, F., Gyoba, J. and Suzuki, Y. (2012). Compression of auditory space during forward self-motion, *PLoS ONE* **7**(6), e39402.

Abstract from the 14th International Multisensory Research Forum, The Hebrew University of Jerusalem, Israel, 2013.

* Contact author: teramoto@csse.muroran-it.ac.jp

 DOI:10.1163/22134808-000S0162

BRILL Multisensory Research 26 Supplement (2013) 216 brill.com/msr

Poster Presentation

Spatial magnitude modulates the perceived duration of oddballs: Evidence against attentional theories of the oddball effect

Devin B. Terhune *, **Benjamin Marshall and Roi Cohen Kadosh**

Department of Experimental Psychology, University of Oxford, UK

Abstract

The oddball effect is the perceived temporal dilation of a deviant stimulus in a train of homogeneous stimuli. A number of models have proposed attentional explanations of the oddball effect, according to which temporal dilation occurs because of the violation of perceptual expectations in a manner akin to repetition suppression effects. On the other hand, the perceived duration of a stimulus is known to depend on its spatial magnitude, such that smaller stimuli are perceived to be shorter in duration. In order to examine these competing effects, we examined the role of the spatial magnitude of oddball stimuli in an oddball duration discrimination task. Participants were presented with a single blue square (oddball) at a random temporal position within a sequence of white squares (standards) of equivalent duration. Oddballs were smaller, equivalent, or larger in physical size and varied in duration. Participants judged whether the oddball was shorter or longer in duration than the standard stimuli. Individual participants' data were fitted with a logistic function using maximum likelihood estimation and the point of subjective equality, corresponding to the oddball duration perceived to be equivalent to the standard duration, was estimated. The perceived duration of oddballs varied as a function of spatial magnitude: relative to the standard stimuli, larger oddballs were perceived to be longer in duration whereas small and equivalent oddballs were perceived to be shorter in duration. These results are inconsistent with attentional models of the oddball effect and warrant revisions of these models.

Keywords

Time perception, spatial magnitude, attention

Abstract from the 14th International Multisensory Research Forum, The Hebrew University of Jerusalem, Israel, 2013.

* Contact author: devin.terhune@psy.ox.ac.uk

 DOI:10.1163/22134808-000S0163

BRILL Multisensory Research 26 Supplement (2013) 217 brill.com/msr

Poster Presentation

The neural bases of cross-modal correspondences: Reality or wishful thinking?

Antonia Thelen [1,2,3,*], **Sarah Amrein** [1] **and Micah M. Murray** [1,4,5]

[1] FENL — Functional Electrical Neuroimaging Laboratory, Switzerland
[2] CHUV — Centre Hospitalier Universitaire Vaudois, Switzerland
[3] UNIL — Université de Lausanne, Switzerland
[4] Radiology Department, CHUV — Centre Hospitalier Universitaire Vaudois, Switzerland
[5] Electroencephalography Brain Mapping Core, Center for Biomedical Imaging (CIBM) of Lausanne and Geneva, Switzerland

Abstract

There has recently been resurgent interest in the notion of cross-modal correspondences — i.e. stimulus features that may be preferentially integrated. The extent to which any such correspondences emanate from intrinsic anatomical connectivity or instead from learned (presumably statistical) regularities in the environment remains unresolved. Anatomical data from non-human primates would suggest that high-frequency auditory representations together with peripheral visual field representations in primary cortices are a preferred locus of low-level integration, whereas functional studies in humans have repeatedly demonstrated effects with centrally-presented stimuli and a range of auditory pitches/bandwidths. The present psychophysics and EEG study examined whether auditory–visual integration systematically varies with acoustic pitch and visual eccentricity. Subjects viewed 2 annuli (either foveally-presented or at 12.5° eccentricity, with surface area controlled for cortical magnification) and indicated which, if either, changed its brightness. The paradigm followed a 3 × 3 within subject design: (no, foveal, or peripheral brightness change) × (no, 500 Hz, or 4000 Hz pure tone presentation; the latter of which were controlled for perceived loudness). Accuracy data were analyzed according to signal detection theory, using sensitivity (d′). Reaction time data were analyzed after dividing by detection rates, using the multisensory response enhancement metric (MRE; see Rach *et al.*, 2011, *Psychological Research*). Preliminary data suggest there to be a main effect of pitch (i.e. larger d′ and MRE when the multisensory conditions included a high pitch vs. low pitch sound). There was no evidence of a main effect of eccentricity or interaction between factors. Ongoing EEG analyses will likewise be discussed.

Keywords

Auditory–visual, crossmodal correspondence, EEG

Abstract from the 14th International Multisensory Research Forum, The Hebrew University of Jerusalem, Israel, 2013.

* Contact author: Antonia.Thelen@chuv.ch

 DOI:10.1163/22134808-000S0164

BRILL Multisensory Research 26 Supplement (2013) 218 brill.com/msr

Poster Presentation

Emotional valence modulates object-related audiovisual processing in the human brain

Alexandra Tinnermann [1,*], Marcus J. Naumer [1], Jasper J. F. van den Bosch [1], Jochen Kaiser [2] and Grit Hein [3]

[1] Crossmodal Object Imaging Lab, Institute of Medical Psychology, Goethe-University Frankfurt am Main, Germany

[2] Institute of Medical Psychology, Goethe-University Frankfurt am Main, Germany

[3] Department of Economics, University of Zurich, Switzerland

Abstract

The processing of audiovisual information is an important aspect of object recognition since many objects are characterized by visual and auditory features. The images and sounds of everyday life objects are often associated with positive or negative emotions. There is evidence that emotional valence modulates the neural processing of object images and sounds presented in isolation. However, the impact of emotional valence on the neural processing of audiovisual objects is unknown. The aim of the present study was to investigate the influence of evaluative conditioning on cortical plasticity in human audiovisual object processing. We used functional magnetic resonance imaging (fMRI) to analyze changes in cortical audiovisual processing of unfamiliar artificial objects and sounds in naive subjects (Pre fMRI) and after a behavioral evaluative conditioning session in which subjects acquired positive and negative emotional associations with the artificial stimuli (Post fMRI). Results from the Post fMRI showed that additional brain regions were recruited for the processing of positively and negatively conditioned stimuli compared to the Pre fMRI data. The processing of positive stimuli additionally involved the left caudate nucleus and the right medial prefrontal cortex whereas the processing of negative stimuli additionally recruited the left insula. Furthermore, a congruence effect for matching emotional stimulus pairs was found in the right transversal temporal gyrus. These results indicate that emotional valence modulates object-related audiovisual processing. Moreover, we were able to show that emotional valence can be induced by a behavioral evaluative conditioning procedure leading to plastic changes in the processing of audiovisual stimuli.

Keywords

fMRI, audiovisual integration, valence, evaluative conditioning

Abstract from the 14th International Multisensory Research Forum, The Hebrew University of Jerusalem, Israel, 2013.

* Contact author: atinnermann85@gmail.com

 DOI:10.1163/22134808-000S0165

BRILL Multisensory Research 26 Supplement (2013) 219–220 brill.com/msr

Poster Presentation

Motor commands induce time compression for tactile stimuli

Alice Tomassini [1,*], Monica Gori [1], Gabriel Baud-Bovy [1], Giulio Sandini [1] and Maria Concetta Morrone [2]

[1] Istituto Italiano di Tecnologia, Genoa, Italy
[2] Department of Physiological Sciences, Università di Pisa; Scientific Institute Stella Maris, Pisa, Italy

Abstract

Saccades cause compression of visual space towards the saccadic target, and also a compression of time, both phenomena thought to be related to the problem of maintaining saccadic stability. Interestingly, similar phenomena occur around the time of hand movements, when brief tactile stimuli are systematically mislocalized in the direction of the movement. In this study we measured whether hand movements also cause an alteration of the perceived timing of the touch signals. Participants compared the temporal separation between two pairs of tactile taps delivered to either the right moving or left stationary hand. An auditory tone cued participants to move their right hand as fast as possible along a track. At variable delays from sound presentation the first pair of tactile taps (test) was presented with a fixed onset asynchrony of 150 ms. After 2 s from test presentation the second pair of taps was delivered (probe) with a variable temporal separation. When the tactile stimuli were presented on the motor effector, their perceived temporal separation was reduced. The time compression began approximately 150 ms before movement onset, it increased until just before the hand started moving and continued during the movement. The time compression was effector-specific, as perceived time was veridical for the left stationary hand, ruling out high-level attentional modulation. The results indicate that tactile time intervals are compressed around the time of hand movements. As for vision the mislocalizations of time and space of touch stimuli may be consequences of a mechanism attempting to achieve perceptual stability.

Keywords

Time perception, touch, action

Abstract from the 14th International Multisensory Research Forum, The Hebrew University of Jerusalem, Israel, 2013.

* Contact author: alice.tomassini@iit.it

 DOI:10.1163/22134808-000S0166

References

Dassonville, P. (1995). Haptic localization and the internal representation of the hand in space, *Experimental Brain Research (Experimentelle Hirnforschung Experimentation Cerebrale)* **106**, 434–448.

Maij, F., de Grave, D. D., Brenner, E. and Smeets, J. B. (2011). Misjudging where you felt a light switch in a dark room, *Experimental Brain Research* **213**, 223–227.

Morrone, M. C., Ross, J. and Burr, D. (2005). Saccadic eye movements cause compression of time as well as space, *Nature Neuroscience* **8**, 950–954.

Watanabe, J., Nakatani, M., Ando, H. and Tachi, S. (2009). Haptic localizations for onset and offset of vibro-tactile stimuli are dissociated, *Experimental Brain Research* **193**, 483–489.

BRILL Multisensory Research 26 Supplement (2013) 221 brill.com/msr

Poster Presentation

Multisensory integration in ferret auditory cortex: Effects of inactivating visual cortex

Stephen Town *, **Katherine Wood and Jennifer Bizley**
University College London, UK

Abstract

Neurons in ferret auditory cortex respond to simple visual stimuli such as an LED and auditory responses to white noise can be modulated by the presence of such visual stimuli (Bizley *et al.*, 2007). Visual influences on auditory cortical neurons may result from input connections from a variety of areas — including visual cortex, parietal cortex and the thalamus. Here, we asked whether visual activity in auditory cortex could be abolished by inactivating visual cortex.

Multi-unit neural activity was recorded at 13 sites within the auditory cortex of three ferrets before and during inactivation of the visual cortex using cooling loops cooled to between 4 and 7°C. We found numerous auditory cortical units in which cooling significantly reduced visual but not auditory responses (measured as mean firing rate). Furthermore units were found in which visual modulation of auditory responses was reduced by cooling. Both findings support the involvement of visual cortex in multisensory integration within auditory cortex.

We also found evidence for the involvement of regions outside of visual cortex: Several units demonstrated visual responses robust to cooling or in which cooling reduced visual responses but only within a circumscribed period of the response. Additionally in rarer cases, cooling led to the emergence of auditory and visual responses or the visual modulation of auditory responses within auditory cortex. Such findings suggest that visual activity in auditory cortex may result from connections to multiple brain regions and that communication from visual to auditory cortex may involve both excitatory and inhibitory mechanisms.

Keywords

Auditory cortex, visual cortex, cortical inactivation

Reference

Bizley, J. K., Nodal, F. R., Bajo, V. M., Nelken, I. and King, A. J. (2007). Physiological and anatomical evidence for multisensory interactions in auditory cortex, *Cereb. Cortex* **17**(9), 2172–2189.

Abstract from the 14th International Multisensory Research Forum, The Hebrew University of Jerusalem, Israel, 2013.

* Contact author: s.town@ucl.ac.uk

 DOI:10.1163/22134808-000S0167

BRILL Multisensory Research 26 Supplement (2013) 222 brill.com/msr

Poster Presentation

Orthographic dependency in the neural correlates of reading: Evidence from audiovisual integration in English readers

Nienke M. van Atteveldt [1,2,*], Ian D. Holloway [2], Leo Blomert [3] and Daniel Ansari [4]

[1] Netherlands Institute for Neuroscience, Amsterdam, The Netherlands
[2] Maastricht University, The Netherlands
[3] The University of Western Ontario, London, ON, Canada
[4] The University of Western Ontario, Canada

Abstract

Adequately learning to read and write is indispensable in our technological societies; however, it is an effortful process that requires years of formal education. In transparent alphabetic orthographies, such as Dutch, the crucial building blocks for literacy skills are the correspondences between letters and speech sounds (LS pairs). Previously, we demonstrated that LS pairs reliably produce a congruency-sensitive fMRI response in the superior temporal cortex (STC) of Dutch fluent readers, but not in dyslexic readers. Here, we used fMRI to investigate whether a similar congruency-sensitivity exists in STC of readers of opaque orthographies, such as English, where the relation between letters and speech sounds is unreliable, and may therefore not provide a similarly important basis for literacy. Eighteen subjects passively perceived congruent and incongruent audiovisual pairs that have different transparencies in English: letters/speech sounds (LS; irregular), letters/letter names (LN; fairly transparent) and numerals/number names (NN; transparent). We detected a congruency by pair-type interaction in STC bilaterally, that was produced by a strong congruency effect for NN, a weaker one for LN, and an incongruency effect for LS. No congruency effects for LS pairs were found in the predicted direction (congruent > incongruent), in contrast to the previous results in Dutch readers. These results indicate that through education, the STC becomes tuned to the congruency of transparent audiovisual pairs, but suggest different neural processing of irregular pairs. This orthographic dependency of LS integration has important implications for remediation strategies for dyslexia in different languages.

Keywords

Reading, orthographic consistency, education, fMRI, audiovisual, cross-linguistic

Abstract from the 14th International Multisensory Research Forum, The Hebrew University of Jerusalem, Israel, 2013.

* Contact author: n.vanatteveldt@maastrichtuniversity.nl

DOI:10.1163/22134808-000S0168

BRILL Multisensory Research 26 Supplement (2013) 223–224 brill.com/msr

Poster Presentation

Exogenous auditory cues decrease audiovisual integration of simple lights and sounds

Nathan van der Stoep [1,*], Marieke A. Janssen [1], Stefan van der Stigchel [1] and Tanja C. W. Nijboer [2]

[1] Utrecht University, Department of Experimental Psychology, The Netherlands
[2] University Medical Center Utrecht and Rehabilitation Center De Hoogstraat, Rudolf Magnus Institute of Neuroscience and Center of Excellence for Rehabilitation Medicine, The Netherlands

Abstract

Both bottom-up attention and multisensory integration are processes that can decrease detection times of simple lights and sounds (Spence, 2010; Stevenson *et al.*, 2012). In previous studies, the interaction between top-down attention and audiovisual integration has been extensively investigated (for a review see Talsma *et al.*, 2010). Yet, little attention has been given to the interaction between bottom-up attention and audiovisual integration. We examined the influence of exogenous auditory cues on the integration of lights and sounds in a simple detection task. We presented valid (cue same side as target), invalid (cue opposite side of target), or no cues at random SOAs (175–250 ms) before the presentation of an auditory, visual, or audiovisual target on the left or right side of the screen. As expected, auditory cues decreased response times (RTs) to auditory, visual, and audiovisual targets, compared to the no cue condition, regardless of their validity. RTs to audiovisual targets were faster compared to RTs to auditory targets only in the no cue condition. To investigate whether audiovisual integration was influenced by bottom-up attention, we compared the cumulative density functions (CDF) of audiovisual RTs and the race-model in each cue condition (Ulrich *et al.*, 2007). We found significant violations of the race-model only in the no cue condition. These results indicate that bottom-up attention facilitates the processing of audiovisual information, but at the same time reduces audiovisual integration. We suggest that bottom-up and top-down forms of attention may influence audiovisual integration in different ways.

Keywords

Attention, multisensory, integration, exogenous cues

Abstract from the 14th International Multisensory Research Forum, The Hebrew University of Jerusalem, Israel, 2013.

* Contact author: n.vanderstoep@uu.nl

DOI:10.1163/22134808-000S0169

References

Spence, C. (2010). Crossmodal spatial attention, *Annals of the New York Academy of Sciences* **1191**, 182–200.

Stevenson, R. A., Krueger Fister, J., Barnett, Z. P., Nidiffer, A. R. and Wallace, M. T. (2012). Interactions between the spatial and temporal stimulus factors that influence multisensory integration in human performance, *Experimental Brain Research* **219**, 121–137.

Talsma, D., Senkowski, D., Soto-Faraco, S. and Woldorff, M. G. (2010). The multifaceted interplay between attention and multisensory integration, *Trends in Cognitive Sciences* **14**, 400–410.

Ulrich, R., Miller, J. and Schröter, H. (2007). Testing the race model inequality: An algorithm and computer programs, *Behavior Research Methods* **29**(2), 291–302.

BRILL Multisensory Research 26 Supplement (2013) 225 brill.com/msr

Poster Presentation

The detection of tactile stimuli while observing another in pain in chronic pain patients and controls

Sophie Vandenbroucke [1,*], **Liesbet Goubert** [1] **and Vanessa Harrar** [2]

[1] Ghent University, Belgium
[2] University of Oxford, UK

Abstract

Introduction: There is evidence that viewing touch or pain can modulate the experience of tactile stimulation. The aim of this study was to investigate whether the observation of needle pricks to another person's hand facilitates the detection of tactile stimuli applied to the hand. We hypothesized this would be different for a group of chronic pain patients compared to controls.

Method: Participants with fibromyalgia ($n = 39$) and controls ($n = 38$) were presented a series of videos showing hands being pricked (e.g., by needles) or control videos (e.g., a sponge being pricked), whilst receiving subtle tactile stimuli themselves in the same spatial location (congruent trials) or on the opposite hand (incongruent trials) as the visual stimuli. Participants were asked to detect the presence and location of the tactile stimulus. Signal detection theory was used to compare whether sensitivity was different for congruent and incongruent trials, and between the groups of participants.

Results: Perceptual sensitivity (d') was significantly higher when videos contained a painful situation, compared to the control videos, and was higher when the videos were presented on the same side of the body as the tactile stimulus (congruency effect). The congruency effect was larger when a painful situation was shown compared to control videos. No difference in sensitivity was found between fibromyalgia patients and controls.

Discussion: This study suggests that the detection of somatic sensations can be facilitated by observing painful visual stimuli. The visual-somatosensory modulation was independent upon the presence of chronic pain.

Keywords

Vicarious sensations, somatosensory modulation, observation of pain

Abstract from the 14th International Multisensory Research Forum, The Hebrew University of Jerusalem, Israel, 2013.

* Contact author: Sophie.Vandenbroucke@ugent.be

 DOI:10.1163/22134808-000S0170

BRILL Multisensory Research 26 Supplement (2013) 226 brill.com/msr

Poster Presentation

Children do not recalibrate motor-sensory temporal order after exposure to delayed sensory feedback

Tiziana Vercillo [1,*], **David C. Burr** [2], **Giulio Sandini** [3] **and Monica Gori** [3]

[1] Robotics, Brain & Cognitive Sciences Department, Istituto Italiano di Tecnologia, Genova, Italy

[2] Department of Neuroscience, University of Florence, Florence, Italy

[3] Istituto Italiano di Tecnologia, Genova, Italy

Abstract

Prolonged adaptation to delayed sensory feedback to a simple motor act (such as pressing a key) seems to provoke a recalibration of sensory-motor synchronization, causing instantaneous feedback to appear to precede the motor act that caused it (Stetson *et al.*, *Neuron*, 2006). We investigated whether a similar recalibration occurs in school-age children, when plasticity may be expected to be even greater than in adults. Subjects adapted to delayed feedback for 100 trials, intermittently pressing a key that caused a tone to sound after a 200 ms delay. During the test phase, subjects responded to a light cue by pressing a key, which was accompanied by a tone played at a variable interval before or after the keypress. Subjects judged whether the tone preceded or followed the keypress. The data were fit with cumulative gaussian psychometric functions, yielding estimates of the points of perceived synchrony (mean) and precision (width). In agreement with previous studies, adults showed a shift in perceived synchrony after adaptation, so the keypress appeared to trail the auditory feedback, implying sensory-motor recalibration. However, school children of all ages (7, 8, 10 & 12 years) showed no adaptation of perceived simultaneity, although their precision in the simultaneity task was comparable with that of adults. This suggests that school-age children do not recalibrate sensory-motor simultaneity after adaptation, implying that the neural mechanisms underlying recalibration require the maturity of both the motor and sensory systems.

Keywords

Temporal recalibration, motor sensory synchrony, development

Abstract from the 14th International Multisensory Research Forum, The Hebrew University of Jerusalem, Israel, 2013.

[*] Contact author: tiziana.vercillo@iit.it

 DOI:10.1163/22134808-000S0171

BRILL
Multisensory Research 26 Supplement (2013) 227
brill.com/msr

Poster Presentation

Audiovisual temporal integration: A freezing effect from rhythms with continuously shifting phases

Manuel Vidal *

Institut de Neurosciences de la Timone, CNRS/Aix-Marseille Université, France

---

Abstract

We constantly experience multimodal events for which signals are transmitted and processed at different speeds. Deciding whether these signals are synchronous or not requires temporal flexibility in the sensory integration. Several mechanisms have been described, from recalibration following exposure to constant lags (Fujisaki *et al.*, 2004; Vroomen *et al.*, 2004a) to temporal ventriloquism in which the perceived visual onset is shifted to that of the nearest auditory onset (Vroomen *et al.*, 2004b). While testing recalibration used constant phase-shifts, we designed stimuli in which the auditory and visual rhythms have slightly different frequencies resulting in lags increasing or decreasing constantly. In each trial, participants viewed 7 flashes and heard 5 tone pips presented roughly within the first 5 flashes period. The task was to report whether the rhythm of the last two flashes was faster or slower than the first five. We manipulated the relative rhythm (auditory faster or slower) and final phase (5th flash at 0° or 180°). The temporal frequencies where chosen so that initial and final lags roughly went from opposite to perfect synchronization or *vice-versa*. We found that the perceived visual pace accelerated when captured by a faster auditory rhythm or released from a slower auditory rhythm, and decelerated when captured by a slower auditory rhythm or released from a faster auditory rhythm. Consistent with temporal ventriloquism, we evidenced a temporal freezing effect in which the perceived visual pace successively speeds up and slows down according to the ongoing audiovisual phase shift.

Keywords

Temporal window of integration, audiovisual, ventriloquist effect

References

Fujisaki, W., Shimojo, S., Kashino, M. and Nishida, S. (2004). Recalibration of audiovisual simultaneity, *Nat. Neurosci.* **7**, 773–778.

Vroomen, J., Keetels, M., De Gelder, B. and Bertelson, P. (2004a). Recalibration of temporal order perception by exposure to audio-visual asynchrony, *Brain Res. Cogn. Brain Res.* **22**, 32–35.

Vroomen, J., De Gelder, B. and Vroomen, J. (2004b). Temporal ventriloquism: sound modulates the flash-lag effect, *J. Exp. Psychol. Hum. Percept. Perform.* **30**, 513–518.

Abstract from the 14th International Multisensory Research Forum, The Hebrew University of Jerusalem, Israel, 2013.

* Contact author: manuel.vidal@univ-amu.fr

DOI:10.1163/22134808-000S0172

BRILL Multisensory Research 26 Supplement (2013) 228–229 brill.com/msr

Poster Presentation

The effect of blindfolding on sound localization

Patrice Voss *, **Vanessa Tabry and Robert J. Zatorre**

McGill University, Canada

Abstract

While vision has been shown to play an important role in calibrating the spatial representations of other senses (Knudsen and Knudsen, 1990; Withington *et al.*, 1994), numerous recent reports have suggested that individuals deprived of vision are actually able to develop heightened auditory spatial abilities (Lessard *et al.*, 1998; Voss *et al.*, 2004). However, most such cases have compared the blind to blindfolded sighted individuals, a procedure that might introduce a strong performance bias in that blind individuals, who have had their whole lives to adapt to this condition, whereas sighted individuals might be put at a severe disadvantage when suddenly being asked to localize sounds without visual input. To address this unknown, we compared the sound localization ability of eight sighted individuals with and without blindfold using a 3D sound presentation device in a hemianechoic chamber. We used a 2 × 2 × 2 factorial design, where we compared two vision conditions (blindfold *vs.* non-blindfold), two sound planes (horizontal *vs.* vertical) and two pointing methods (finger *vs.* head). A 2 × 2 × 2 repeated measures ANOVA revealed a significant effect of vision (no-blindfold > blindfold; $p < 0.001$) and a significant effect of plane (horizontal > vertical; $p < 0.001$). Moreover, a vision × plane × pointing triple interaction was also significant ($p = 0.004$), and was primarily driven by a significantly poorer performance of head pointing in the horizontal plane when blindfolded, compared to the non-blindfolded condition. This result argues strongly against the use of head pointing methodologies with blindfolded individuals, particularly in the horizontal plane, as it likely introduces a robust bias when comparing them to blind individuals.

Keywords

Sound localization, visual feedback, blindness

References

Knudsen, E. I. and Knudsen, P. F. (1990). Sensitive and critical periods for visual calibration of sound localization by barn owls, *Journal of Neuroscience* **10**, 222–232.

Withington, D., Binns, K., Keating, M., Thornton, S. and Ingham, N. (1994). Post-crucial period effects of visual experience and deprivation on the guinea-pig superior collicular map of auditory space, *Experimental Physiology* **79**, 313–318.

Abstract from the 14th International Multisensory Research Forum, The Hebrew University of Jerusalem, Israel, 2013.

* Contact author: patrice.voss@mcgill.ca

DOI:10.1163/22134808-000S0173

Lessard, N., Pare, M., Lepore, F. and Lassonde, M. (1998). Early-blind human subjects localize sound sources better than sighted subjects, *Nature* **395**, 278–280.

Voss, P., Lassonde, M., Gougoux, F., Fortin, M., Guillemot, J. P. and Lepore, F. (2004). Early- and late-onset blind individuals show supra-normal auditory abilities in far-space, *Current Biology* **14**, 1734–1738.

BRILL *Multisensory Research 26 Supplement (2013) 231–236* brill.com/msr

Subject Index to Volume 26 Supplement

Abstracts of the 14th International Multisensory Research Forum, The Hebrew University of Jerusalem, Israel, 3rd–6th June 2013

 DOI:10.1163/22134808-000S0175

T

BRILL Multisensory Research 26 Supplement (2013) 237–242 brill.com/msr

Author Index to Volume 26 Supplement

Abstracts of the 14th International Multisensory Research Forum, The Hebrew University of Jerusalem, Israel, 3rd–6th June 2013

 DOI:10.1163/22134808-000S0175

Call for Papers

We are happy to announce there will be a Special Issue published in *Multisensory Research* following the IMRF 2013 conference in Jerusalem. Submissions will be accepted from all who are interested, but preference will be given for speakers participating and presenting their work at the IMRF 2013.

Citation for IMRF special issues has tended to be very high and, based on statistics from previous years, competition for places in the Special Issue is likely to be correspondingly high. Papers will have the format of regular articles in *Multisensory Research* and will be peer reviewed in the normal way. Papers can be submitted through http://www.editorialmanager.com/msrbrill/ and will be handled by the Guest Editors Amir Amedi, Shelly Levy-Tzedek and David I. Shore. Select the 'Special Issue: 14th International Multisensory Research Forum' when submitting.

More details about the special issue will be available soon.

Laurence R. Harris and Concetta Morrone
Editors-in-Chief

See other abstracts online as Supplementary Material

DOI:10.1163/22134808-000S0176

www.ingramcontent.com/pod-product-compliance
Lightning Source LLC
Chambersburg PA
CBHW071437230926
51846CB00001B/268
* 9 7 8 9 0 0 4 2 5 6 3 2 3 *